Practice for Stanford 9

Grade 6

Harcourt Brace & Company
Orlando • Atlanta • Austin • Boston • San Francisco • Chicago • Dallas • New York • Toronto • London
www.hbschool.com

Printed in the United States of America

ISBN 0-15-313753-3

1 2 3 4 5 6 7 8 9 10 073 2000 99

CONTENTS

Overview

Multiple-Choice Practice

Description of Practice for STANFORD 9 Books

The *Practice for STANFORD 9* books will help you prepare students for standardized achievement tests. There is a book for each of grades 1–8, each book providing at least one page of practice items for each objective in the Stanford 9 and three comprehensive practice tests that encompass objectives found on most standardized tests.

The practice tests are provided at three levels of difficulty: a simplified version of a standardized test; a version at the level of difficulty of many standardized tests; and a version at, or slightly above, the level of difficulty of most standardized tests given at this grade level.

Uses of Practice for STANFORD 9 Books

Purpose

These books may be used diagnostically to monitor students' progress, to review previously taught skills, or to prepare for district or state standardized tests.

Using the Practice Pages

Each practice page covers one objective of the Stanford 9. Using the bubble-response answer sheet on page xiii helps students develop skills necessary for taking standardized tests.

The pace at which you use the practice items depends on your particular classes or groups and on your teaching style. You may want to use one of the following ideas or a combination of several ideas.

- Have students complete the page of practice items immediately after the concept is presented in class.
- Group appropriate practice pages and have students complete them at given intervals, such as every two weeks or at the end of each grading period.
- Provide a block of time at the beginning of the second semester or just before district or state standardized tests are administered. Have students complete all the practice items before they try any practice tests.
- Use the practice pages in conjunction with the practice tests as detailed in the following paragraphs.

Using the Practice Tests

A bubble-response answer sheet designed specifically for these tests is on page xiv. Using this form helps students develop skills and confidence that will benefit them in taking standardized tests.

How you use the practice tests depends on your particular class or group. You might want to use one of the following ideas or any appropriate alternative.

- Administer the first (simplified) practice test. Based on the test results, have students complete the appropriate practice pages. Then administer the second test to determine needed remediation. If further remediation is required, refer to the correlation chart beginning on page 97 for exercises in *MATH ADVANTAGE.* Finally, administer the third practice test if necessary.
- At the start of the second semester or shortly before the actual standardized tests will be given, administer the second or third practice test. Use the test results to determine which objectives to teach or review. A review can utilize either the practice pages in this book or exercises from *MATH ADVANTAGE,* which can be easily found in the correlation beginning on page 97. You may want to administer the first practice test to provide students with a successful testing experience just prior to their taking the actual standardized test.

Administering the Practice Tests

The practice tests are organized in two parts—*Problem Solving,* which covers content skills and appropriate applications, and *Procedures,* which covers computation and computation in a problem-solving format. You may want to give students a short break between the two parts of the test.

Prior to administering any of the practice tests, read the following general directions to the students:

We are going to take a test. We will be practicing things we have learned in class. Do your best and try to answer each question. You will need a ruler that measures in inches and in centimeters. You will also need a Number 2 pencil to complete this test.

Be sure that all students have the necessary equipment: pencils, erasers, and rulers. No other materials, such as protractors, compasses, or calculators are permitted.

Distribute the tests and appropriate bubble-response answer sheets (found on page xiv). Have students write their names on the answer sheets. Point out that there are two parts, and give instructions such as one of the following:

There are two parts to this test. When you finish the first part, go on to the second part. Be sure you are recording your answers in the proper spaces.

There are two parts to this test. When you finish the Problem Solving section, STOP. Do not go on.

Scoring

An answer key for the practice pages is on pages viii-xi of this book. You may wish to score these practice pages yourself or have students help in scoring them. It is recommended that all items students missed be discussed individually, in small groups, or by the entire class.

An answer key for each practice test begins on page xii of this book. The method of scoring these tests is left to the discretion of the teacher. For other suggestions, see *Using the Practice Tests* on the previous page of this book.

Grade 6 Answer Keys

Objective 1
Page 1

Sample	C
1.	D
2.	H
3.	A
4.	G
5.	E
6.	J
7.	B
8.	H

Objective 1
Page 2

Sample	A
1.	D
2.	H
3.	C
4.	H
5.	B
6.	K
7.	B
8.	H

Objective 2
Page 3

Sample	A
1.	C
2.	H
3.	B
4.	F
5.	D
6.	H
7.	C
8.	H

Objective 3
Page 4

Sample	B
1.	B
2.	G
3.	B
4.	H
5.	B
6.	F

Objective 3
Page 5

Sample	B
1.	C
2.	G
3.	D
4.	G
5.	C
6.	H

Objective 4
Page 6

Sample	B
1.	B
2.	G
3.	D
4.	J
5.	B
6.	G

Objective 5
Page 7

Sample	D
1.	A
2.	K
3.	A
4.	F
5.	C
6.	H
7.	B
8.	K
9.	C

Objective 6
Page 8

Sample	B
1.	B
2.	K
3.	D
4.	H
5.	A
6.	H
7.	D
8.	J
9.	A
10.	G

Objective 7
Page 9

Sample	D
1.	E
2.	J
3.	C
4.	H
5.	C
6.	G
7.	B
8.	H

Objective 7
Page 10

Sample	C
1.	C
2.	F
3.	E
4.	J
5.	B
6.	K
7.	B
8.	H

Objective 8
Page 11

Sample	B
1.	A
2.	G
3.	C
4.	H
5.	E
6.	H
7.	B
8.	H
9.	A
10.	G
11.	B

Objective 8
Page 12

Sample	C
1.	B
2.	H
3.	B
4.	H
5.	A
6.	H
7.	E
8.	J
9.	A
10.	G

Objective 8
Page 13

Sample	E
1.	A
2.	J
3.	A
4.	K
5.	B
6.	F
7.	C

Objective 9
Page 14

Sample	A
1.	E
2.	F
3.	C
4.	F
5.	D
6.	K
7	C
8.	G
9.	C

Objective 10
Page 15

Sample	C
1.	A
2.	G
3.	E
4.	J
5.	A
6.	J
7.	C
8.	H

Objective 11
Page 16

Sample	C
1.	A
2.	G
3.	B
4.	H
5.	C
6.	K
7.	C

Objective 12
Page 17

Sample	B
1.	E
2.	H
3.	A
4.	F
5.	C
6.	J
7.	E
8.	G

Objective 13
Page 18

Sample	D
1.	B
2.	H
3.	B
4.	K
5.	C
6.	G
7	A

Objective 14
Page 19

Sample	C
1.	D
2.	F
3.	C
4.	G

Objective 15
Page 20

Sample	C
1.	B
2.	H
3.	C
4.	J
5.	B

Objective 16
Page 21

Sample	C
1.	B
2.	J
3.	A
4.	J
5.	C

Objective 17
Page 22

Sample	B
1.	D
2.	J
3.	B
4.	G
5.	D

Objective 17
Page 23

Sample	B
1.	B
2.	J
3.	C
4.	G
5.	C

Objective 18
Page 24

Sample	B
1.	A
2.	J
3.	D
4.	H
5.	C

Objective 18
Page 25

Sample	B
1.	D
2.	G
3.	D
4.	G
5.	A
6.	F
7.	D
8.	J
9.	A
10.	H

Objective 19
Page 26

Sample	C
1.	B
2.	H
3.	D
4.	J
5.	B
6.	G
7.	B
8.	J

Objective 20
Page 27

Sample	C
1.	B
2.	H
3.	C
4.	G
5.	C
6.	F
7.	D

Objective 21
Page 28

Sample	B
1.	A
2.	J
3.	C
4.	H
5.	A
6.	J
7.	B
8.	H
9.	B
10.	G
11.	C

Objective 22
Page 29

Sample	D
1.	A
2.	F
3.	A
4.	H
5.	D
6.	H
7.	D

Objective 22
Page 30

Sample	D
1.	B
2.	H
3.	D
4.	F
5.	A
6.	J

Objective 23
Page 31

Sample	D
1.	D
2.	G
3.	A
4.	G
5.	D
6.	H

Objective 23
Page 32

Sample	B
1.	C
2.	F
3.	C
4.	G
5.	A

Objective 24
Page 33

Sample	B
1.	B
2.	H
3.	A
4.	J
5.	C
6.	F

Objective 24
Page 34

Sample	A
1.	C
2.	H
3.	C
4.	J
5.	D

Objective 25
Page 35

Sample	C
1.	A
2.	H
3.	C
4.	G

Objective 25
Page 36

Sample	D
1.	A
2.	J
3.	C
4.	F
5.	A

Objective 26
Page 37

Sample	A
1.	D
2.	F
3.	A
4.	J
5.	B
6.	J
7.	D

Objective 27
Page 38

Sample	D
1.	D
2.	H
3.	B
4.	F
5.	C

Objective 28
Page 39

Sample	C
1.	A
2.	G
3.	D
4.	J

Objective 29
Page 40

Sample	A
1.	D
2.	H
3.	A
4.	J

Objective 30
Page 41

Sample	A
1.	D
2.	H
3.	B
4.	J

Objective 30
Page 42

Sample	C
1.	A
2.	J
3.	A

Objective 31
Page 43

Sample	B
1.	D
2.	J
3.	B
4.	J
5.	B

Objective 32
Page 44

Sample	B
1.	D
2.	G
3.	C
4.	G
5.	D

Objective 32
Page 45

Sample	D
1.	A
2.	H
3.	B
4.	H
5.	A
6.	G

Objective 33
Page 46

Sample	B
1.	B
2.	F
3.	B
4.	G
5.	A
6.	H
7.	B
8.	H
9.	D

Objective 33
Page 47

Sample	A
1.	C
2.	H
3.	B
4.	H
5.	A
6.	H
7.	B
8.	H
9.	B
10.	J
11.	D

Objective 33
Page 48

Sample	A
1.	B
2.	G
3.	D
4.	H
5.	B
6.	H
7.	A
8.	J
9.	B
10.	H
11.	D

Objective 33
Page 49

Sample	A
1.	C
2.	G
3.	B
4.	F
5.	A
6.	H
7.	C
8.	H
9.	A
10.	F
11.	B

Objective 34
Page 50

Sample	C
1.	A
2.	G
3.	A
4.	G
5.	E
6.	H
7.	A
8.	J
9.	C

Objective 34
Page 51

Sample	C
1.	C
2.	F
3.	C
4.	H
5.	E
6.	F
7.	C
8.	J
9.	A

Objective 35
Page 52

Sample	C
1.	B
2.	H
3.	B
4.	J
5.	E
6.	G
7.	D
8.	G
9.	A

Objective 36
Page 53

Sample	B
1.	C
2.	F
3.	C
4.	G
5.	B
6.	H

Objective 37
Page 54

Sample	A
1.	D
2.	F
3.	B
4.	H
5.	B
6.	H

Objective 37
Page 55

Sample	C
1.	B
2.	J
3.	B
4.	F
5.	C
6.	F

Objective 38
Page 56

Sample	A
1.	D
2.	G
3.	C
4.	J
5.	B
6.	H

Objective 38
Page 57

Sample	A
1.	C
2.	J
3.	B
4.	J
5.	C
6.	J

Objective 39
Page 58

Sample	C
1.	C
2.	G
3.	B
4.	H
5.	C

Objective 39
Page 59

Sample	C
1.	B
2.	H
3.	C
4.	J
5.	D

Objective 40
Page 60

Sample	C
1.	C
2.	F
3.	D
4.	H
5.	B
6.	G

Objective 41
Page 61

Sample	C
1.	B
2.	J
3.	B

Objective 41
Page 62

Sample	D
1.	A
2.	F
3.	A

Objective 41
Page 63

Sample	B
1.	D
2.	H
3.	A
4.	G
5.	A

Objective 42
Page 64

Sample	B
1.	C
2.	F
3.	D
4.	H
5.	D
6.	H
7.	A

Objective 42
Page 65

Sample	C
1.	C
2.	H
3.	A
4.	J
5.	C
6.	G

Objective 43
Page 66

Sample	A
1.	D
2.	H
3.	A
4.	H
5.	C
6.	J

Objective 44
Page 67

Sample	D
1.	A
2.	G
3.	D
4.	J
5.	A
6.	G
7.	B

Objective 45
Page 68

Sample	D
1.	C
2.	H
3.	C
4.	J
5.	B

Objective 46
Page 69

Sample	B
1.	B
2.	G
3.	A
4.	J
5.	C
6.	G

Objective 46
Page 70

Sample	A
1.	C
2.	H
3.	C
4.	J
5.	D
6.	G

Objective 46
Page 71

Sample	B
1.	B
2.	F
3.	C
4.	G
5.	D
6.	F

Objective 46
Page 72

Sample	D
1.	C
2.	H
3.	A
4.	H
5.	C

Objective 47
Page 73

Sample	B
1.	C
2.	F
3.	B
4.	H
5.	A
6.	H
7.	D
8.	F
9.	C

Objective 48
Page 74

Sample	D
1.	A
2.	G
3.	C
4.	G
5.	C
6.	F
7.	C

Objective 48
Page 75

Sample	B
1.	A
2.	J
3.	A
4.	H
5.	C
6.	H
7.	D

Objective 49
Page 76

Sample	C
1.	D
2.	G
3.	B
4.	F
5.	C

Objective 50
Page 77

Sample	D
1.	A
2.	F
3.	B
4.	F
5.	C
6.	J
7.	B
8.	J

Objective 50
Page 78

Sample	A
1.	B
2.	H
3.	C
4.	H
5.	A
6.	H
7.	D

Practice Test 1
Pages 79-83

Part 1
Problem Solving

Sample	D
1	C
2	G
3	A
4	F
5	B
6	H
7	C
8	H
9	D
10	J
11	A
12	G
13	C
14	G
15	B
16	H
17	B
18	G
19	C
20	H
21	B
22	J
23	B
24	F
25	C
26	G

Part 2
Procedures

Sample	B
1	E
2	J
3	C
4	J
5	C
6	H
7	E
8	H
9	B
10	H
11	A
12	F
13	A
14	K
15	D

Practice Test 2
Pages 85-89

Part 1
Problem Solving

Sample	B
1	A
2	J
3	A
4	H
5	C
6	H
7	B
8	G
9	A
10	H
11	D
12	H
13	C
14	J
15	C
16	H
17	A
18	J
19	D
20	J
21	C
22	H
23	D
24	H
25	C
26	H

Part 2
Procedures

Sample	C
1	C
2	G
3	E
4	G
5	D
6	K
7	A
8	F
9	A
10	J
11	A
12	K
13	D
14	H
15	A

Practice Test 3
Pages 91-95

Part 1
Problem Solving

Sample	A
1	C
2	G
3	A
4	J
5	D
6	J
7	D
8	H
9	D
10	H
11	D
12	J
13	B
14	J
15	D
16	H
17	A
18	F
19	A
20	H
21	B
22	G
23	D
24	F
25	C
26	H

Part 2
Procedures

Sample	C
1	D
2	J
3	E
4	G
5	E
6	H
7	A
8	F
9	C
10	J
11	A
12	J
13	E
14	J
15	C

Name ______________________________ Date ______________

Practice Page Answer Sheet

MATH ADVANTAGE

Practice Page ____________

For each question, fill in the circle for the correct answer. Remember, you can only mark E and K as answers for the practice pages that have a "Not Here" answer.

SAMPLE

1. Ⓐ Ⓑ Ⓒ Ⓓ Ⓔ
2. Ⓕ Ⓖ Ⓗ Ⓙ Ⓚ
3. Ⓐ Ⓑ Ⓒ Ⓓ Ⓔ
4. Ⓕ Ⓖ Ⓗ Ⓙ Ⓚ
5. Ⓐ Ⓑ Ⓒ Ⓓ Ⓔ
6. Ⓕ Ⓖ Ⓗ Ⓙ Ⓚ
7. Ⓐ Ⓑ Ⓒ Ⓓ Ⓔ
8. Ⓕ Ⓖ Ⓗ Ⓙ Ⓚ
9. Ⓐ Ⓑ Ⓒ Ⓓ Ⓔ
10. Ⓕ Ⓖ Ⓗ Ⓙ Ⓚ
11. Ⓐ Ⓑ Ⓒ Ⓓ Ⓔ
12. Ⓕ Ⓖ Ⓗ Ⓙ Ⓚ

Name ______________________________ Date ______________

Practice Test Answer Sheet

MATH ADVANTAGE

Mathematics: Problem Solving

Practice Test ________

SAMPLE
Ⓐ Ⓑ Ⓒ Ⓓ Ⓔ

1. Ⓐ Ⓑ Ⓒ Ⓓ Ⓔ
2. Ⓕ Ⓖ Ⓗ Ⓙ Ⓚ
3. Ⓐ Ⓑ Ⓒ Ⓓ Ⓔ
4. Ⓕ Ⓖ Ⓗ Ⓙ Ⓚ
5. Ⓐ Ⓑ Ⓒ Ⓓ Ⓔ
6. Ⓕ Ⓖ Ⓗ Ⓙ Ⓚ
7. Ⓐ Ⓑ Ⓒ Ⓓ Ⓔ
8. Ⓕ Ⓖ Ⓗ Ⓙ Ⓚ
9. Ⓐ Ⓑ Ⓒ Ⓓ Ⓔ
10. Ⓕ Ⓖ Ⓗ Ⓙ Ⓚ
11. Ⓐ Ⓑ Ⓒ Ⓓ Ⓔ
12. Ⓕ Ⓖ Ⓗ Ⓙ Ⓚ
13. Ⓐ Ⓑ Ⓒ Ⓓ Ⓔ
14. Ⓕ Ⓖ Ⓗ Ⓙ Ⓚ
15. Ⓐ Ⓑ Ⓒ Ⓓ Ⓔ
16. Ⓕ Ⓖ Ⓗ Ⓙ Ⓚ
17. Ⓐ Ⓑ Ⓒ Ⓓ Ⓔ
18. Ⓕ Ⓖ Ⓗ Ⓙ Ⓚ
19. Ⓐ Ⓑ Ⓒ Ⓓ Ⓔ
20. Ⓕ Ⓖ Ⓗ Ⓙ Ⓚ
21. Ⓐ Ⓑ Ⓒ Ⓓ Ⓔ
22. Ⓕ Ⓖ Ⓗ Ⓙ Ⓚ
23. Ⓐ Ⓑ Ⓒ Ⓓ Ⓔ
24. Ⓕ Ⓖ Ⓗ Ⓙ Ⓚ
25. Ⓐ Ⓑ Ⓒ Ⓓ Ⓔ
26. Ⓕ Ⓖ Ⓗ Ⓙ Ⓚ

Mathematics: Procedures

SAMPLE
Ⓐ Ⓑ Ⓒ Ⓓ Ⓔ

1. Ⓐ Ⓑ Ⓒ Ⓓ Ⓔ
2. Ⓕ Ⓖ Ⓗ Ⓙ Ⓚ
3. Ⓐ Ⓑ Ⓒ Ⓓ Ⓔ
4. Ⓕ Ⓖ Ⓗ Ⓙ Ⓚ
5. Ⓐ Ⓑ Ⓒ Ⓓ Ⓔ
6. Ⓕ Ⓖ Ⓗ Ⓙ Ⓚ
7. Ⓐ Ⓑ Ⓒ Ⓓ Ⓔ

Name ______________________________

Practice Objective 1

DIRECTIONS

Read each question and choose the best answer. Then mark the space for the answer you have chosen. If a correct answer is *not here,* mark the space for NH.

SAMPLE

What is 482 rounded to the nearer hundred?

A 400 **C** 500 **E** NH

B 480 **D** 1,000

1 **The attendance at a football game was 19,853. What is that number rounded to the nearer thousand?**

A 10,000 **C** 19,900 **E** NH

B 19,000 **D** 20,000

2 **The population of Germany is 83,536,115. What is that number rounded to the nearer million?**

F 90,000,000

G 86,000,000

H 84,000,000

J 80,000,000

K NH

3 **What is the standard form of eighty million, seven hundred five thousand, forty-three?**

A 80,705,043

B 80,750,043

C 8,705,043

D 80,705,430

E NH

4 **What is the value of 4 in the number 2,947,850?**

F Four thousands

G Four ten thousands

H Four millions

J Four hundreds

K NH

5 **Which number is less than 2,780?**

A 2,870 **C** 3,078 **E** NH

B 2,980 **D** 8,002

6 **Which number is greater than 3,507?**

F 3,490 **H** 2,509 **K** NH

G 3,315 **J** 3,705

7 **Which shows the numbers written from greatest to least?**

A 3,665; 3,566; 3,656; 6,356

B 6,356; 3,665; 3,656; 3,566

C 3,566; 3,656; 3,665; 6,356

D 6,356; 3,566; 3,665; 3,656

E NH

8 **Which shows the numbers written from least to greatest?**

F 6,540; 6,504; 5,460; 4,605

G 4,605; 6,504; 6,540; 5,460

H 4,605; 5,460; 6,504; 6,540

J 4,605; 6,504; 5,460; 6,540

K NH

Name ______________________________

Practice Objective 1

DIRECTIONS

Read each question and choose the best answer. Then mark the space for the answer you have chosen. If a correct answer is *not here,* mark the space for NH.

SAMPLE

What is $7.42 rounded to the nearer dollar?

A $7 C $8 E NH
B $7.40 D $10

1 **What is $73.59 rounded to the nearer dollar?**

A $70 C $73.60 E NH
B $73 D $74

2 **Flo Joyner holds the world record for the 100-meter dash. She finished in 10.49 seconds. What is that number rounded to the nearer tenth?**

F 10 sec H 10.5 sec K NH
G 10.4 sec J 11 sec

3 **The average speed of the winning car in the second Indy 500 race was 78.719 miles per hour. What is that number rounded to the nearer hundredth?**

A 79 C 78.72 E NH
B 78.71 D 78.7

4 **What is the standard form of fifty thousand, twenty-eight and seventy-one thousandths?**

F 50,280.71 J 50,028.0071
G 50,028.71 K NH
H 50,028.071

5 **What is the value of 8 in 73.085?**

A 8 tens
B 8 hundredths
C 8 thousandths
D 8 tenths
E NH

6 **Which number is greater than 5.709?**

F 0.795 H 5.7 K NH
G 5.09 J 5.099

7 **Which shows the numbers written from greatest to least?**

A 5.28, 8.06, 7.53, 8.43
B 8.43, 8.06, 7.53, 5.28
C 8.43, 7.53, 5.28, 8.06
D 5.28, 7.53, 8.06, 8.43
E NH

8 **Which shows the numbers written from least to greatest?**

F 5.602, 5.026, 5.062, 6.502
G 6.502, 5.602, 5.062, 5.026
H 5.026, 5.062, 5.602, 6.502
J 5.026, 5.062, 6.502, 5.602
K NH

Name ______________________________

DIRECTIONS

Read each question and choose the best answer. Then mark the space for the answer you have chosen. If a correct answer is *not here,* mark the space for NH.

SAMPLE

Which decimal names the same number as 30 + 5 + 0.7 + 0.01?

A 35.71 C 305.71 E NH
B 35.071 D 305.071

1 Which decimal names the same number as 40 + 3 + 0.5?

A 43.05 C 43.5 E NH
B 403.05 D 403.5

2 Which decimal names the same number as 300 + 70 + 6 + 0.2 + 0.06?

F 376.026 H 376.26 K NH
G 376.206 J 3,076.26

3 Which decimal names the same number as 900 + 5 + 0.7 + 0.01?

A 950.071 C 905.701 E NH
B 905.71 D 905.0701

4 Which decimal names the same number as 2,000 + 60 + 3 + 0.8?

F 2,063.8 H 2,063.008 K NH
G 2,063.08 J 2,603.008

5 Which gives 324.809 in expanded form?

A 300 + 20 + 4 + 0.08 + 0.9
B 300 + 20 + 4 + 0.8 + 0.09
C 300 + 20 + 4 + 0.08 + 0.009
D 300 + 20 + 4 + 0.8 + 0.009
E NH

6 Which gives 407.092 in expanded form?

F 400 + 7 + 0.9 + 0.002
G 400 + 70 + 0.09 + 0.002
H 400 + 7 + 0.09 + 0.002
J 40 + 7 + 0.9 + 0.02
K NH

7 Which gives 23,050.407 in expanded form?

A 20,000 + 3,000 + 50 + 0.04 + 0.7
B 20,000 + 3,000 + 50 + 0.4 + 0.07
C 20,000 + 3,000 + 50 + 0.4 + 0.007
D 20,000 + 3,000 + 50 + 0.04 + 0.007
E NH

8 Which gives 45,006.019 in expanded form?

F 40,000 + 500 + 6 + 0.01 + 0.009
G 40,000 + 5,000 + 600 + 0.1 + 0.09
H 40,000 + 5,000 + 6 + 0.01 + 0.009
J 4,000 + 500 + 6 + 0.1 + 0.09
K NH

Name ______________________________

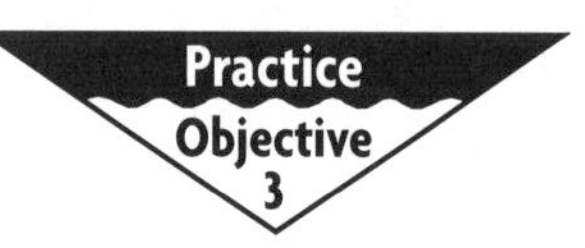

DIRECTIONS

Read each question and choose the best answer. Then mark the space for the answer you have chosen. If a correct answer is *not here*, mark the space for NH.

SAMPLE

Carol made these bank deposits in December: $89, $191, and $92. *About* how much did she deposit in December?

A About $480

B About $400

C About $300

D About $100

1 Carmen traveled 3,949 miles on vacation. Geraldine traveled 1,258 miles. *About* how much farther did Carmen travel than Geraldine?

A About 1,000 miles

B About 3,000 miles

C About 4,000 miles

D About 5,000 miles

2 There are 32 watercoolers in an office building. Each watercooler holds 26 gallons of water. *About* how many gallons are there in all?

F About 60 gallons

G About 900 gallons

H About 6,000 gallons

J About 9,000 gallons

3 The librarians put 3,281 books on shelves at the library. They put 27 books on each shelf. *About* how many shelves did they need?

A About 90 shelves

B About 100 shelves

C About 150 shelves

D About 200 shelves

4 Sasha read one book with 325 pages, another with 549 pages, and another with 908 pages. *About* how many pages did she read?

F About 3,000 pages

G About 2,500 pages

H About 1,700 pages

J About 1,300 pages

5 There are 5,356 books in the school library. Of these books, 4,219 can be loaned to students. *About* how many books cannot be loaned to students?

A About 2,000 books

B About 1,000 books

C About 800 books

D About 500 books

6 A glacier can move about 225 inches in 74 years. *About* how far can it move in one year?

F About 3 inches

G About 30 inches

H About 300 inches

J About 3,000 inches

Name ______________________________

DIRECTIONS

Read each question and choose the best answer. Then mark the space for the answer you have chosen. If a correct answer is *not here,* mark the space for NH.

SAMPLE

A large department store has 65 western hats, 42 straw hats, 71 baseball caps, and 36 dressy hats for sale. *About* how many hats did they have?

A About 150 hats
B About 200 hats
C About 250 hats
D About 300 hats

1 **Neka has 128 blue marbles, 36 green marbles, 72 red marbles, and 63 yellow marbles. *About* how many marbles does she have in all?**

A About 200 marbles
B About 250 marbles
C About 300 marbles
D About 150 marbles

2 **In June, 250 people went to the park. In July, 850 people went to the park, and 750 people went to the park in August. *About* how many people went to the park during these months?**

F About 1,600 people
G About 1,850 people
H About 2,000 people
J About 2,850 people

3 **The Rodriguez family traveled 3,538 miles in 57 hours. *About* how many miles did they travel each hour?**

A About 240,000 miles
B About 6,000 miles
C About 600 miles
D About 60 miles

4 **Twenty-two buses were needed to take 1,276 fans to the basketball playoffs. *About* how many people rode in each bus?**

F About 50 people
G About 60 people
H About 500 people
J About 600 people

5 **Darnell scored these points on four quizzes. *About* how many points did he score in all?**
41 36 42 44

A About 120 points
B About 140 points
C About 160 points
D About 200 points

6 **Keisha ran these distances in three days. *About* how far did she run altogether?**
815 yards 790 yards 792 yards

F About 1,800 yards
G About 2,100 yards
H About 2,400 yards
J About 3,600 yards

Name ______________________________

Practice Objective 4

DIRECTIONS

Read each question and choose the best answer. Then mark the space for the answer you have chosen. If a correct answer is *not here,* mark the space for NH.

SAMPLE

Jimmy bought 3.9 pounds of peanuts, 1.9 pounds of potato chips, and 4.8 pounds of pretzels. How many pounds of snack food did he buy?

Estimate. Then decide which answer is most reasonable.

A 12.8 pounds
B 10.6 pounds
C 9.3 pounds
D 8.7 pounds

1 **During each daily gym class, Jeannie runs around the track 11 times. How many times will she run around the track in a school year with 172 days?**

Estimate. Then decide which answer is most reasonable.

A 1,458 times
B 1,892 times
C 2,845 times
D 3,025 times

2 **Jill clipped 3,356 recipes in 1998 and 2,748 recipes in 1999. Karen clipped a total of 4,219 recipes. How many more recipes did Jill clip than Karen?**

Estimate. Then decide which answer is most reasonable.

F 2,942 recipes
G 1,885 recipes
H 10,221 recipes
J 1,043 recipes

3 **Patti hiked $6\frac{3}{4}$ miles on Saturday and $2\frac{7}{8}$ miles on Sunday. How much farther did she hike on Saturday than on Sunday?**

Estimate. Then decide which answer is most reasonable.

A $9\frac{5}{8}$ miles
B $5\frac{1}{8}$ miles
C $8\frac{3}{8}$ miles
D $3\frac{7}{8}$ miles

4 **At the apple orchard, John picked $13\frac{7}{8}$ pounds of red delicious apples and $9\frac{1}{2}$ pounds of gala apples. How many pounds of apples did he pick all together?**

Estimate. Then decide which answer is most reasonable.

F $4\frac{3}{8}$ lb
G $4\frac{1}{4}$ lb
H $22\frac{3}{4}$ lb
J $23\frac{3}{8}$ lb

5 **Lizzie's bowling score for one game is usually between 80 and 110. What is a reasonable answer for her point total in 8 games?**

A About 550 points
B About 750 points
C About 900 points
D About 1,000 points

6 **Flannel fabric is sale priced at $7.15 for 2 yards. What is a reasonable estimate for the cost of 19 yards of flannel fabric?**

F About $100
G About $70
H About $60
J About $45

Name ______________________________

DIRECTIONS

Read each question and choose the best answer. Then mark the space for the answer you have chosen. If a correct answer is *not here*, mark the space for NH.

SAMPLE

$$\begin{array}{r} 4.5 \\ +\,3.29 \\ \hline \end{array}$$

A 7.29 C 7.709 E NH
B 7.7 D 7.79

1

$$\begin{array}{r} 3.29 \\ +\,7.48 \\ \hline \end{array}$$

A 10.77 C 11.77 E NH
B 10.67 D 1,177

2 **61.87 + 3.2 + 7.09**

F 7,216 H 69.28 K NH
G 71.16 J 16.477

3

$$\begin{array}{r} 20.31 \\ -\,18.64 \\ \hline \end{array}$$

A 1.67 C 12.57 E NH
B 10.67 D 38.95

4 **57.3 − 38.65**

F 18.65 H 32.92 K NH
G 29.75 J 95.95

5 **6 − 1.38**

A 7.38 C 4.62 E NH
B 5.38 D 4.38

6

$$\begin{array}{r} 6.2 \\ \times\,2.5 \\ \hline \end{array}$$

F 1,550 H 15.5 K NH
G 155 J 15.4

7 **0.71 × 3.8**

A 0.2698 C 26.98 E NH
B 2.698 D 269.8

8

$$\begin{array}{r} 2.05 \\ \times\,1.66 \\ \hline \end{array}$$

F 34,030 H 34.03 K NH
G 340.3 J 0.3403

9 **Sally buys 3.4 pounds of ground beef, 2.8 pounds of ground pork, and 1.7 pounds of ground veal. How many pounds of ground meat does she buy?**

A 6 pounds
B 6.9 pounds
C 7.9 pounds
D 8 pounds
E NH

Name ______________________________

Practice Objective 6

DIRECTIONS

Read each question and choose the best answer. Then mark the space for the answer you have chosen. If a correct answer is *not here,* mark the space for NH.

SAMPLE

4)26.8

A 0.67 C 67 E NH
B 6.7 D 6.2

1 **8)108.8**

A 1.36 C 136 E NH
B 13.6 D 11.1

2 **6)16.02**

F 3.67 H 2.6 K NH
G 20.67 J 267

3 **18)52.2**

A 290 C 3.2 E NH
B 29 D 2.9

4 **0.6)2.28**

F 38 H 3.8 K NH
G 0.38 J 3.08

5 **2.3)9.384**

A 4.08 C 4.8 E NH
B 480 D 408

6 **1.4)2.086**

F 14.9 H 1.49 K NH
G 149 J 0.149

7 **3.3)27.192**

A 0.824 C 824 E NH
B 82.4 D 8.24

8 **4.9)22.589**

F 0.0461 H 0.0461 K NH
G 461 J 4.61

9 **Jim and three friends equally share $3.68. How much money does each person get?**

A $ 0.92
B $14.72
C $9.20
D $11.04
E NH

10 **Cindy has a piece of lumber that is 14.4 meters long. She wants to make shelves for her room. Each shelf will be 2.4 meters long. How many shelves can she cut from the piece of lumber?**

F 5 shelves
G 6 shelves
H 7 shelves
J 35 shelves
K NH

Name ____________________

Practice Objective 7

DIRECTIONS

Read each question and choose the best answer. Then mark the space for the answer you have chosen. If a correct answer is *not here,* mark the space for NH.

SAMPLE

Which fraction is greater than $\frac{1}{3}$?

A $\frac{1}{8}$ C $\frac{2}{6}$ E NH

B $\frac{2}{9}$ D $\frac{2}{3}$

1 Which fraction is greater than $\frac{4}{9}$?

A $\frac{1}{4}$ C $\frac{8}{18}$ E NH

B $\frac{1}{18}$ D $\frac{4}{10}$

2 Which fraction is less than $\frac{2}{3}$?

F $\frac{10}{15}$ H $\frac{7}{10}$ K NH

G $\frac{3}{4}$ J $\frac{3}{5}$

3 Which mixed number is greater than $2\frac{3}{4}$?

A $2\frac{4}{9}$ C $2\frac{9}{10}$ E NH

B $2\frac{3}{5}$ D $1\frac{4}{5}$

4 Which mixed number is less than $20\frac{1}{10}$?

F $20\frac{9}{10}$ H $20\frac{1}{20}$ K NH

G $30\frac{1}{100}$ J $20\frac{1}{5}$

5 Which shows the fractions written from least to greatest?

A $\frac{1}{2}, \frac{1}{3}, \frac{1}{4}, \frac{1}{10}$ D $\frac{1}{2}, \frac{1}{3}, \frac{1}{4}, \frac{1}{10}$

B $\frac{1}{4}, \frac{1}{3}, \frac{1}{2}, \frac{1}{10}$ E NH

C $\frac{1}{10}, \frac{1}{4}, \frac{1}{3}, \frac{1}{2}$

6 Which shows the fractions written from greatest to least?

F $\frac{9}{10}, \frac{3}{5}, \frac{3}{10}, \frac{1}{2}$ J $\frac{1}{2}, \frac{3}{5}, \frac{3}{10}, \frac{9}{10}$

G $\frac{9}{10}, \frac{3}{5}, \frac{1}{2}, \frac{3}{10}$ K NH

H $\frac{3}{10}, \frac{1}{2}, \frac{3}{5}, \frac{9}{10}$

7 Which is the greatest mixed number?
$2\frac{5}{8}, 2\frac{3}{4}, 2\frac{1}{2}, 2\frac{7}{16}, 2\frac{1}{4}$

A $2\frac{5}{8}$ C $2\frac{1}{2}$ E $2\frac{1}{4}$

B $2\frac{3}{4}$ D $2\frac{7}{16}$

8 Which shows the mixed numbers written from greatest to least?

F $12\frac{3}{4}, 13\frac{1}{2}, 12\frac{1}{3}, 13\frac{2}{5}$

G $13\frac{2}{5}, 13\frac{1}{2}, 12\frac{3}{4}, 12\frac{1}{3}$

H $13\frac{1}{2}, 13\frac{2}{5}, 12\frac{3}{4}, 12\frac{1}{3}$

J $12\frac{1}{3}, 12\frac{3}{4}, 13\frac{2}{5}, 13\frac{1}{2}$

K NH

Name ______________________________

DIRECTIONS

Read each question and choose the best answer. Then mark the space for the answer you have chosen. If a correct answer is *not here,* mark the space for NH.

SAMPLE

Which fraction is equivalent to $\frac{1}{4}$?

A $\frac{2}{10}$ C $\frac{2}{8}$ E NH

B $\frac{4}{12}$ D $\frac{8}{16}$

1 Which fraction is equivalent to $\frac{3}{8}$?

A $\frac{6}{10}$ C $\frac{6}{16}$ E NH

B $\frac{8}{16}$ D $\frac{6}{24}$

2 What is the missing number?
$4\frac{7}{10} = \frac{n}{10}$

F 47 H 38 K NH

G 41 J 21

3 What is the missing number?
$8\frac{3}{4} = \frac{n}{4}$

A 15 C 27 E NH

B 20 D 29

4 Which fraction is equivalent to $\frac{12}{20}$?

F $\frac{12}{40}$ H $\frac{24}{20}$ K NH

G $\frac{24}{30}$ J $\frac{3}{5}$

5 Emmy has finished $\frac{9}{12}$ of a rug she is making. Which fraction is the simplest form of $\frac{9}{12}$?

A $\frac{2}{3}$ C $\frac{18}{24}$ E NH

B $\frac{3}{4}$ D $\frac{12}{16}$

6 The model shows how much pizza Kirk had left for his party after he ate a slice. Which fraction could you use to show this?

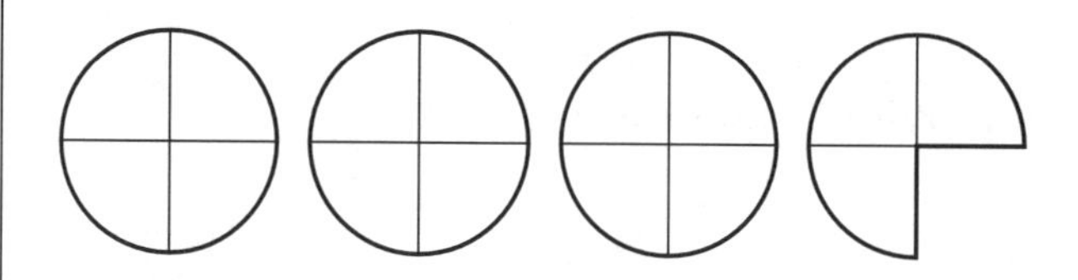

F $4\frac{1}{4}$ H $3\frac{1}{4}$ K NH

G $\frac{4}{15}$ J $\frac{15}{8}$

7 Mary has $3\frac{1}{4}$ loaves of bread. Which fraction is another name for $3\frac{1}{4}$?

A $\frac{13}{3}$ C $\frac{7}{4}$ E NH

B $\frac{13}{4}$ D $\frac{4}{13}$

8 Kyle has $\frac{25}{8}$ pizza. What is $\frac{25}{8}$ written as a mixed number?

F $2\frac{1}{8}$ H $3\frac{1}{8}$ K NH

G $4\frac{1}{8}$ J $8\frac{1}{3}$

Name ______________________________

Practice Objective 8

DIRECTIONS

Read each question and choose the best answer. Then mark the space for the answer you have chosen. If a correct answer is *not here,* mark the space for NH.

SAMPLE

What is the GCF (greatest common factor) of 12 and 16?

A 2 **C** 6 **E** NH

B 4 **D** 8

1 **What is the GCF of 9 and 12?**

A 3 **C** 9 **E** NH

B 4 **D** 12

2 **What is the GCF of 28 and 16?**

F 2 **H** 8 **K** NH

G 4 **J** 16

3 **What is the GCF of 32 and 24?**

A 2 **C** 8 **E** NH

B 4 **D** 16

4 **What is the GCF of 21 and 35?**

F 3 **H** 7 **K** NH

G 5 **J** 21

5 **Choose the simplest form for $\frac{12}{20}$.**

A $\frac{3}{10}$ **C** $\frac{2}{3}$ **E** NH

B $\frac{6}{10}$ **D** $3\frac{1}{3}$

6 **Choose the simplest form for $\frac{15}{35}$.**

F $\frac{3}{8}$ **H** $\frac{3}{7}$ **K** NH

G $\frac{1}{3}$ **J** $\frac{5}{7}$

7 **Choose the simplest form for $\frac{48}{64}$.**

A $\frac{2}{3}$ **C** $\frac{6}{8}$ **E** NH

B $\frac{3}{4}$ **D** $\frac{12}{16}$

8 **Choose the simplest form for $\frac{24}{32}$.**

F $\frac{3}{5}$ **H** $\frac{3}{4}$ **K** NH

G $\frac{2}{3}$ **J** $\frac{5}{6}$

9 **Choose the simplest form for $\frac{35}{40}$.**

A $\frac{7}{8}$ **C** $\frac{5}{7}$ **E** NH

B $\frac{4}{5}$ **D** $\frac{7}{9}$

10 **Choose the simplest form for $\frac{36}{48}$.**

F $\frac{2}{3}$ **H** $\frac{12}{16}$ **K** NH

G $\frac{3}{4}$ **J** $\frac{18}{24}$

11 **Choose the simplest form for $\frac{32}{56}$.**

A $\frac{1}{2}$ **C** $\frac{16}{28}$ **E** NH

B $\frac{4}{7}$ **D** $\frac{8}{14}$

Name ______________________________

Practice Objective 8

DIRECTIONS

Read each question and choose the best answer. Then mark the space for the answer you have chosen. If a correct answer is *not here,* mark the space for NH.

SAMPLE

What is the LCM (least common multiple) of 4 and 6?

A 4   C 12   E NH
B 6   D 24

1 **What is the LCM of 2 and 3?**

A 3   C 12   E NH
B 6   D 24

2 **What is the LCM of 8 and 6?**

F 16   H 24   K NH
G 18   J 36

3 **Which pair of numbers has 24 as the LCM?**

A 2, 12   C 3, 6   E NH
B 3, 8   D 24, 28

4 **The denominators of $\frac{3}{8}$ and $\frac{1}{3}$ are 8 and 3. What is their LCM?**

F 12   H 24   K NH
G 18   J 48

5 **The denominators of $\frac{3}{8}$ and $\frac{5}{16}$ are 8 and 16. What is their LCM?**

A 16   C 4   E NH
B 8   D 2

6 **The denominators of $\frac{2}{3}$ and $\frac{3}{10}$ are 3 and 10. What is their LCM?**

F 3   H 30   K NH
G 10   J 60

7 **The denominators of $\frac{11}{15}$ and $\frac{2}{5}$ are 15 and 5. What is their LCM?**

A 5   C 20   E NH
B 10   D 25

8 **Find $1\frac{7}{8} - \frac{3}{4}$.**

F $\frac{1}{8}$ H 1   K NH
G $2\frac{5}{8}$ J $1\frac{1}{8}$

9 **Find $2\frac{3}{16} - \frac{2}{3}$.**

A $1\frac{25}{48}$ C $2\frac{23}{48}$ E NH
B $1\frac{1}{2}$ D $2\frac{41}{48}$

10 **Find $2\frac{11}{12} + 1\frac{7}{8}$.**

F $1\frac{19}{24}$ H $1\frac{1}{24}$ K NH
G $4\frac{19}{24}$ J $3\frac{19}{24}$

Name ______________________________

DIRECTIONS

Read each question and choose the best answer. Then mark the space for the answer you have chosen. If a correct answer is *not here,* mark the space for NH.

SAMPLE

$$\frac{1}{2} + \frac{2}{3}$$

A $\frac{2}{5}$ C $\frac{1}{6}$ E NH

B $\frac{5}{6}$ D $\frac{2}{3}$

1

$$\frac{1}{8} + \frac{1}{4}$$

A $\frac{3}{8}$ C $\frac{3}{12}$ E NH

B $\frac{1}{6}$ D $\frac{3}{16}$

2

$$9\frac{7}{8} + 1\frac{2}{3}$$

F $10\frac{13}{24}$ H $11\frac{1}{8}$ K NH

G $1\frac{13}{24}$ J $11\frac{13}{24}$

3

$$\frac{5}{8} - \frac{1}{4}$$

A $\frac{3}{8}$ C $\frac{1}{2}$ E NH

B $\frac{7}{8}$ D $\frac{1}{8}$

4

$$8\frac{7}{16} - 3\frac{1}{4}$$

F $11\frac{11}{16}$ H $5\frac{3}{8}$ K NH

G $5\frac{11}{16}$ J $4\frac{3}{16}$

5

$$15\frac{11}{12} - 3\frac{5}{8}$$

A $19\frac{13}{24}$ C $13\frac{13}{24}$ E NH

B $12\frac{7}{24}$ D $12\frac{1}{3}$

6 Carrie buys $3\frac{1}{2}$ lb of red potatoes and $4\frac{1}{8}$ lb of Idaho potatoes. How many pounds of potatoes does she buy in all?

F $7\frac{5}{8}$ lb H 8 lb K NH

G $7\frac{1}{4}$ lb J $\frac{5}{8}$ lb

7 Darl runs $1\frac{3}{5}$ mile. Cassie runs $2\frac{1}{5}$ miles. How much farther does Cassie run than Darl?

A $3\frac{4}{5}$ mi C $\frac{3}{5}$ mi E NH

B $1\frac{2}{5}$ mi D 1 mi

Name ______________________

Practice Objective 9

DIRECTIONS

Read each question and choose the best answer. Then mark the space for the answer you have chosen. If a correct answer is *not here,* mark the space for NH.

SAMPLE

$\frac{1}{2} \times \frac{5}{8}$

A $\frac{5}{16}$ C $\frac{5}{10}$ E NH

B $\frac{10}{16}$ D $\frac{6}{10}$

1 $\frac{2}{5} \times \frac{3}{8}$

A $\frac{6}{20}$ C $6\frac{2}{3}$ E NH

B $\frac{5}{40}$ D $\frac{3}{40}$

2 $6 \times 4\frac{1}{3}$

F 26 H $25\frac{1}{3}$ K NH

G 25 J 24

3 $2\frac{1}{2} \times 3\frac{2}{5}$

A $9\frac{1}{2}$ C $8\frac{1}{2}$ E NH

B 9 D $6\frac{1}{5}$

4 $\frac{1}{2} \div 6$

F $\frac{1}{12}$ H 3 K NH

G $\frac{1}{3}$ J 12

5 $\frac{2}{3} \div \frac{8}{15}$

A $\frac{4}{5}$ C $1\frac{1}{5}$ E NH

B $\frac{16}{45}$ D $1\frac{1}{4}$

6 $2\frac{1}{3} \div \frac{1}{2}$

F $\frac{3}{14}$ H $1\frac{1}{6}$ K NH

G $\frac{6}{7}$ J $5\frac{1}{3}$

7 **Mel had 32 sets of baseball cards. He sold $\frac{3}{8}$ of them. How many sets of cards did he sell?**

A 4 sets C 12 sets E NH

B 9 sets D 16 sets

8 **Carla's recipe calls for $4\frac{1}{2}$ teaspoons of sugar. Since she is making only half the recipe, she only needs $\frac{1}{2}$ of that amount. How much sugar does Carla need to use?**

F $2\frac{1}{8}$ teaspoons

G $2\frac{1}{4}$ teaspoons

H 5 teaspoons

J 9 teaspoons

K NH

9 **Denise has a $10\frac{1}{2}$-inch board. She cuts it into $\frac{3}{4}$-inch pieces. How many pieces does she get?**

A $7\frac{3}{4}$ pieces

B 8 pieces

C 14 pieces

D $14\frac{1}{4}$ pieces

E NH

Name ______________________________

DIRECTIONS

Read each question and choose the best answer. Then mark the space for the answer you have chosen. If a correct answer is *not here,* mark the space for NH.

SAMPLE

Which number is a composite number?

5, 13, 24, 29

A 5 **C** 24 **E** NH

B 13 **D** 29

1 **Which number is a prime number?**

13, 40, 49, 69

A 13 **C** 49 **E** NH

B 40 **D** 69

2 **Which number is a composite number?**

17, 27, 53, 31

F 17 **H** 31 **K** NH

G 27 **J** 53

3 **Which number is a prime number?**

15, 21, 39, 55

A 15 **C** 39 **E** NH

B 21 **D** 55

4 **Which number is a composite number?**

3, 5, 17, 21

F 3 **H** 17 **K** NH

G 5 **J** 21

5 **Which list shows all the prime numbers between 30 and 40?**

A 31, 37

B 31, 33, 37, 39

C 31, 35

D 33, 37

E 31, 33, 35, 37, 39

6 **Which list shows all the prime numbers less than 10?**

F 1, 3, 5, 7, 9

G 3, 5 , 7, 9

H 7

J 2, 3, 5, 7

K 3, 5, 7

7 **Which list shows all the prime numbers between 50 and 60?**

A 51, 53, 55, 57, 59

B 51, 53, 59

C 53, 59

D 53, 57, 59

E 53, 57

8 **Which list shows all the prime numbers between 70 and 80?**

F 71, 73, 75, 77, 79

G 71, 73, 77

H 71, 73, 79

J 73, 79

K 75, 77

Name ______________________________

Practice Objective 11

DIRECTIONS

Read each question and choose the best answer. Then mark the space for the answer you have chosen. If a correct answer is *not here,* mark the space for NH.

SAMPLE

Which is the greatest integer?

A $^-2$　　C 3　　E $^-9$

B 0　　D $^-4$

1 **Which is less than 0?**

A $^-3$　　C 0　　E 1

B 2　　D 3

2 **Which is greater than 0?**

F $^-10$　　H $^-9$　　K 0

G $^+4$　　J $^-3$

3 **What is the opposite of 5?**

A 5　　C $|5|$　　E NH

B $^-5$　　D $\frac{1}{5}$

4 **Which of these shows the integers written in order from least to greatest?**

F $^-1, ^-2, 0, 4$

G $4, ^-2, ^-1, 0$

H $^-2, ^-1, 0, 4$

J $0, ^-1, ^-2, 4$

K NH

5 **Which of these shows the integers written in order from greatest to least?**

A $^-8, ^-6, 4, ^-2$

B $^-8, ^-6, ^-2, 4$

C $4, ^-2, ^-6, ^-8$

D $^-8, 4, ^-2, ^-6$

E NH

6 **Marisa's scores for five rounds were $^-9$, $^-8$, 7, 6, and 1. Order the numbers from greatest to least.**

F $^-8, 6, 7, 1, ^-9$

G $^-8, 1, 6, 7, ^-9$

H $^-9, ^-8, 1, 6, 7$

J $7, 6, 1, ^-9, ^-8$

K NH

7 **Ian recorded these low temperatures in degrees Fahrenheit for 5 days.**
4°F, $^-3$°F, 2°F, $^-1$°F, 0°F
Order the numbers from least to greatest.

A $4, ^-3, 2, ^-1, 0$

B $4, 2, 0, ^-1, ^-3$

C $^-3, ^-1, 0, 2, 4$

D $0, ^-1, 2, ^-3, 4$

E $0, ^-1, ^-3, 2, 4$

Name ____________________

DIRECTIONS

Read each question and choose the best answer. Then mark the space for the answer you have chosen. If a correct answer is *not here,* mark the space for NH.

SAMPLE

$^{-}13 + ^{+}7 =$

A $^{-}20$ C 6 E NH

B $^{-}6$ D 20

1 $^{-}10 + ^{+}5 =$

A $^{-}15$ C $^{+}10$ E NH

B $^{+}5$ D $^{+}15$

2 $^{+}8 + ^{-}6 =$

F $^{-}14$ H $^{+}2$ K NH

G $^{-}2$ J $^{+}14$

3 $^{-}24 + ^{-}9 =$

A $^{-}33$ C $^{+}15$ E NH

B $^{-}15$ D $^{+}33$

4 **The temperature was $^{-}5$°C at noon. By sunset, the temperature had dropped 12°. What was the temperature at sunset?**

F $^{-}17$°C H 7°C K NH

G $^{-}7$°C J 17°C

5 $^{+}7 - ^{+}12 =$

A $^{+}19$ C $^{-}5$ E NH

B $^{+}5$ D $^{-}19$

6 $^{-}4 - ^{+}11 =$

F $^{+}15$ H $^{-}7$ K NH

G $^{+}7$ J $^{-}15$

7 $^{+}14 - ^{-}11 =$

A $^{+}23$ C $^{-}3$ E NH

B $^{+}3$ D $^{-}23$

8 **During the football game, the home team lost 4 yards on the first play. On the next play they gained 15 yards. Describe the total change of the home team's position using integers.**

F $^{-}4 + ^{-}15 = ^{-}19$

G $^{-}4 + 15 = ^{+}11$

H $^{+}4 + ^{-}15 = ^{-}11$

J $^{+}4 + 15 = ^{+}19$

K $^{-}4 - ^{+}15 = ^{-}19$

Name ________________________________

DIRECTIONS

Read each question and choose the best answer. Then mark the space for the answer you have chosen. If a correct answer is *not here,* mark the space for NH.

SAMPLE

What is $\frac{3}{4}$ written as a percent?

A 34% C 60% E NH
B 43% D 75%

Use the figure for questions 1–3.

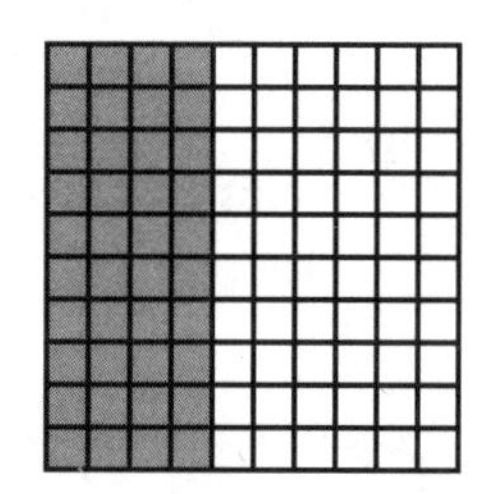

1 **What decimal shows the part of the whole square that is shaded?**

A 0.040 C 0.6 E NH
B 0.40 D 40

2 **What fraction of the figure is shaded? Write the fraction in lowest terms.**

F $\frac{40}{60}$ H $\frac{2}{5}$ K NH
G $\frac{1}{2}$ J $\frac{4}{100}$

3 **What percent of the squares are shaded?**

A 4% C 44% E NH
B 40% D 140%

Use the figure for questions 4–6.

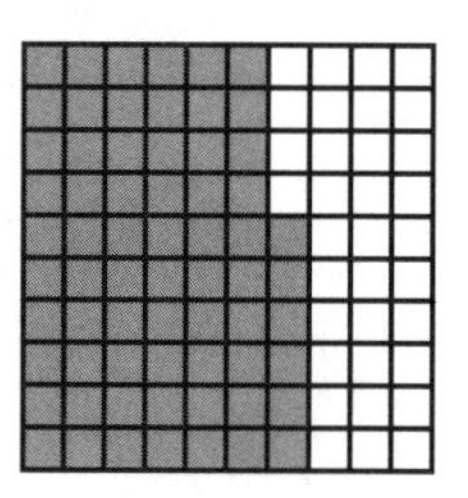

4 **What decimal shows the part of the square that is shaded?**

F 6.6 H 0.34 K NH
G 0.65 J 0.066

5 **What fraction of the figure is shaded? Write the fraction in lowest terms.**

A $\frac{33}{100}$ C $\frac{33}{50}$ E NH
B $\frac{34}{100}$ D $\frac{2}{3}$

6 **What percent of the figure is shaded?**

F 660% H 34% K NH
G 66% J 6.6%

7 **What percent of the tile floor is gray?**

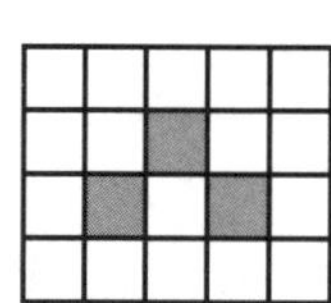

A 15% C 30% E NH
B 20% D $33\frac{1}{3}$%

Name ______________________________

DIRECTIONS

Read each question and choose the best answer. Then mark the space for the answer you have chosen.

SAMPLE

Suppose you bought items for $2.75, $1.50, and $8.75. You gave the clerk a $20 bill. Which calculator key sequence would give you the correct change?

A 20 [−] 2.75 [+] 1.50 [+] 8.75 [=]

B 2.75 [+] 1.50 [+] 8.75 [−] 20 [=]

C 20 [−] 2.75 [−] 1.50 [−] 8.75 [=]

D 20 [−] 2.75 [+] 1.50 [+] 8.75 [=]

1 **A machine fills 8 bottles every 10 seconds. To find how many bottles can be filled in 35 seconds, Jane wrote this proportion: $\frac{8}{10} = \frac{n}{35}$. Which key sequence should she use to solve the proportion?**

A 8 [+] 35 [÷] 10 [=]

B 8 [×] 35 [×] 10 [=]

C 8 [×] 35 [−] 10 [=]

D 8 [×] 35 [÷] 10 [=]

2 **Mel bought a $50 jacket on sale. The discount was 30%. To find the amount of the discount, which key sequence would you use?**

F 50 [×] 0.30 [=]

G 30 [×] 0.05 [=]

H 0.30 [×] 50 [×] 15 [=]

J 0.50 [×] 30 [×] 15 [=]

3 **Lorenzo used a calculator to change $\frac{5}{8}$ to a decimal. What did he do?**

A Divide the denominator by the numerator.

B Multiply the numerator by the denominator.

C Divide the numerator by the denominator.

D Multiply the denominator by the numerator.

4 **On a simple four-function calculator, Dennis used the key strokes**

3 [×] 3 [=] [=] [=] [=]

to find which of the following?

F 9^4

G 3^5

H 9×4

J 3^4

Name ______________________________

DIRECTIONS

Read each question and choose the best answer. Then mark the space for the answer you have chosen.

SAMPLE

Which Venn diagram shows the relationship between the vowels {a, e, i, o, u} and the letters of the alphabet?

A 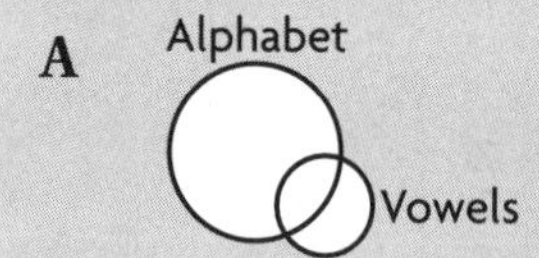

B 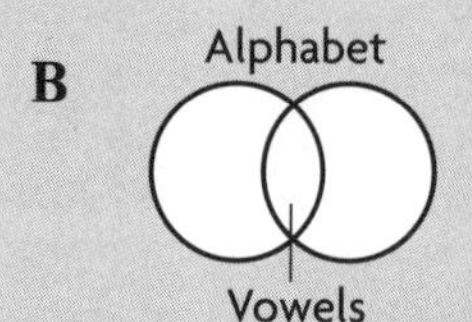

C 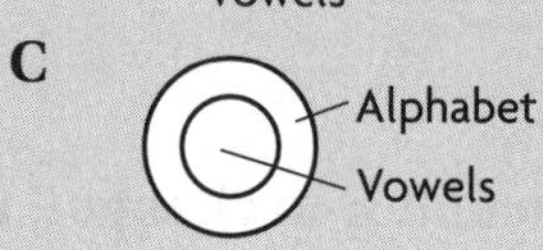

D 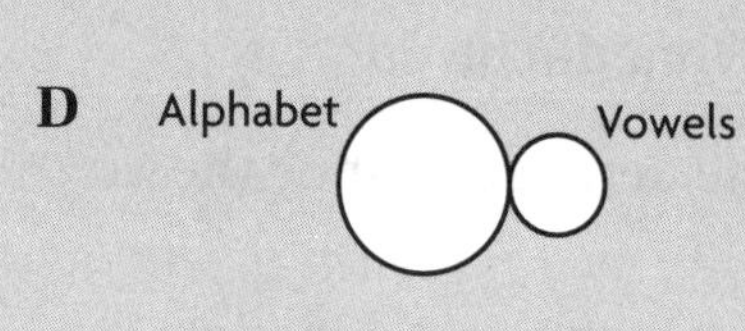

1 What does the Venn diagram show?

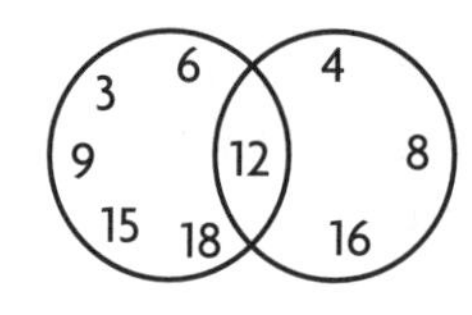

A 12 is a multiple of 4 or a multiple of 3.

B 12 is a common multiple of 3 and 4.

C 16 is a multiple of 3.

D 18 is a multiple of 4.

2 Choose the word that best completes this statement.

{2, 4, 6} is ____?____ of {1, 2, 3, 4, 5, 6}.

F A set

G A member

H A subset

J An element

3 List the elements of the set of all the days of the week.

A {Monday, Tuesday, Wednesday, Thursday, Friday}

B {Saturday,Sunday}

C {Sunday, Monday, Tuesday, Wednesday, Thursday, Friday, Saturday}

D {Sunday, Monday, Tuesday, Wednesday, Thursday, Friday}

4 Describe the set {1, 3, 5, 7, 9}.

F Even numbers less than 10

G Prime numbers less than 10

H Numbers less than 10

J Odd numbers less than 10

5 In which sets does the number $^{-}50$ belong?

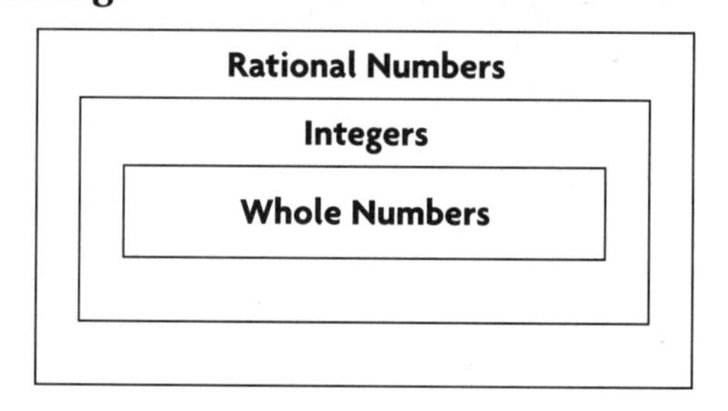

A The sets of whole numbers, integers, and rational numbers

B The sets of integers and rational numbers

C The set of rational numbers

D The set of whole numbers

Name ______________________________

Practice Objective 16

DIRECTIONS

Read each question and choose the best answer. Then mark the space for the answer you have chosen.

Use the Venn diagram showing the numbers of students studying a foreign language for the sample question and question 1.

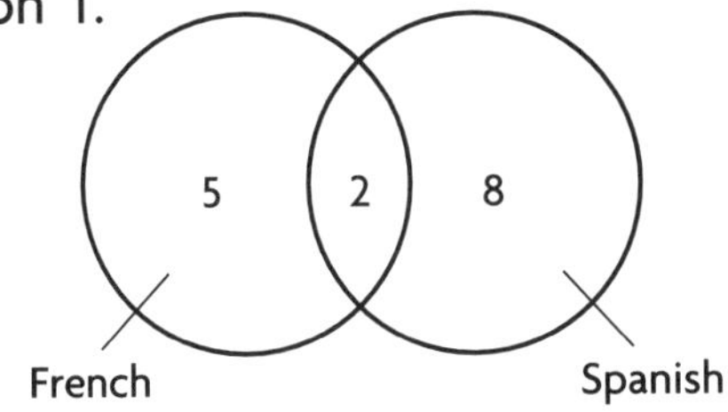

SAMPLE

What does the "2" in the diagram tell you?

A Two students study Spanish.

B Two students study French.

C Two students study both Spanish and French.

D Two students do not study Spanish or French.

1 **How many students are studying only Spanish?**

A 15 students **C** 5 students

B 8 students **D** 2 students

2 **How many students play both the piano and the organ?**

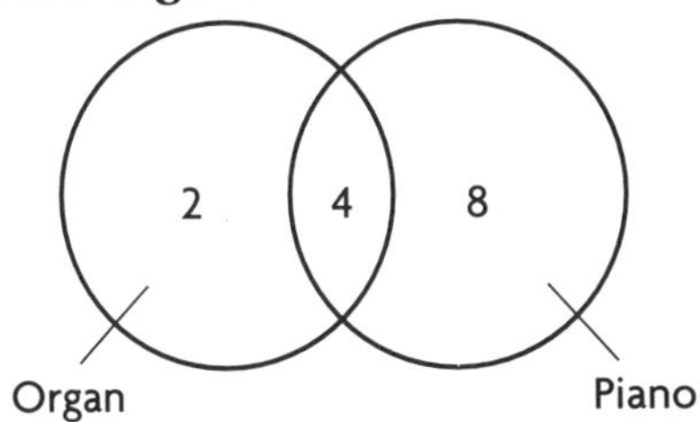

F 12 students **H** 8 students

G 10 students **J** 4 students

3 **How many students play only baseball?**

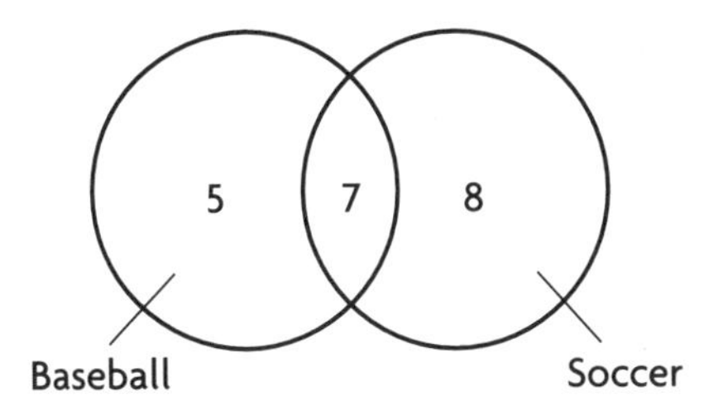

A 5 students **C** 8 students

B 7 students **D** 12 students

4 **How many players are on one or both of these teams?**

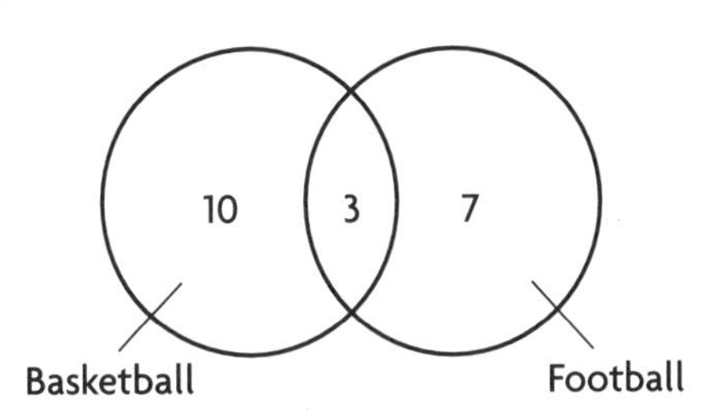

F 7 students **H** 8 students

G 12 students **J** 20 students

5 **Which numbers from the set {1, 3, 4, 5, 6, 8, 9, 11} belong in the shaded section of the diagram?**

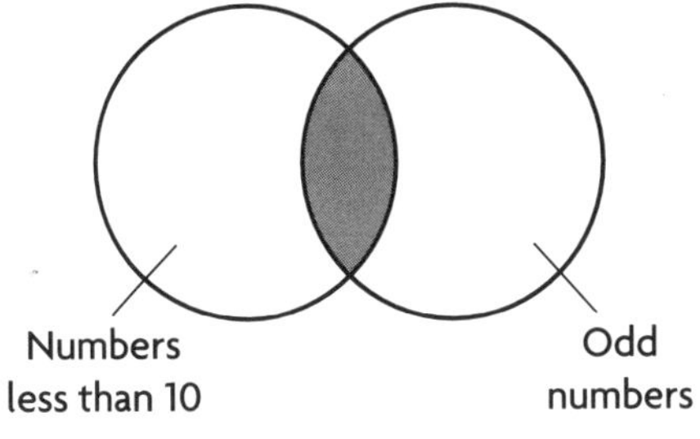

A {11}

B {4, 6, 8}

C {1, 3, 5, 9}

D {1, 3, 4, 5, 6, 8, 9}

Name ____________________

Practice Objective 17

DIRECTIONS

Read each question and choose the best answer. Then mark the space for the answer you have chosen.

SAMPLE

Kari is shading squares on her hundreds chart according to a number pattern.

1	2	3	4	5	6	7	8	9	10
11	12	13	14	15	16	17	18	19	20
21	22	23	24	25	26	27	28	29	30
31	32	33	34	35	36	37	38	39	40
41	42	43	44	45	46	47	48	49	50
51	52	53	54	55	56	57	58	59	60
61	62	63	64	65	66	67	68	69	70
71	72	73	74	75	76	77	78	79	80
81	82	83	84	85	86	87	88	89	90
91	92	93	94	95	96	97	98	99	100

Which number should she shade next to continue the pattern?

A 52   C 61

B 60   D 71

1 **Minette is shading squares on her hundreds chart according to a number pattern.**

1	2	3	4	5	6	7	8	9	10
11	12	13	14	15	16	17	18	19	20
21	22	23	24	25	26	27	28	29	30
31	32	33	34	35	36	37	38	39	40
41	42	43	44	45	46	47	48	49	50
51	52	53	54	55	56	57	58	59	60
61	62	63	64	65	66	67	68	69	70
71	72	73	74	75	76	77	78	79	80
81	82	83	84	85	86	87	88	89	90
91	92	93	94	95	96	97	98	99	100

Which number should she shade next to continue the pattern?

A 71   C 80

B 75   D 81

2 **Carly is working a puzzle that uses this number pattern.**

90, 81, 72, 63, ■

What is the missing number?

F 27   H 45

G 36   J 54

3 **Denzel is working a puzzle that uses this number pattern.**

1.6, 3.3, 5.0, 6.7, ■

What is the missing number?

A 7.4   C 9.4

B 8.4   D 9.7

4 **Norm is working a puzzle that uses this number pattern.**

$\frac{1}{27}, \frac{1}{9}, \frac{1}{3}, 1,$ ■

What is the missing number?

F 9   H $2\frac{2}{3}$

G 3   J $1\frac{1}{3}$

5 **Gavin is working a puzzle that uses this number pattern.**

3, 8, 5, 10, 7, 12, ■, ■, ■

What is the last number in this puzzle?

A 18   C 13

B 15   D 11

Name ______________________________

DIRECTIONS

Read each question and choose the best answer. Then mark the space for the answer you have chosen.

SAMPLE

How many cubes would you use to make the next figure in this pattern?

A 10 cubes
B 12 cubes
C 14 cubes
D 18 cubes

1 **How many cubes would you use to make the next figure in this pattern?**

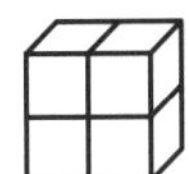
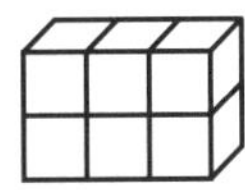

A 3×2
B 4×2
C 5×2
D 6×2

2 **What figure comes next in this pattern?**

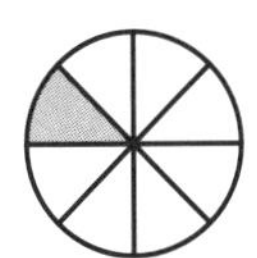
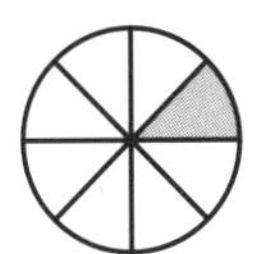

F 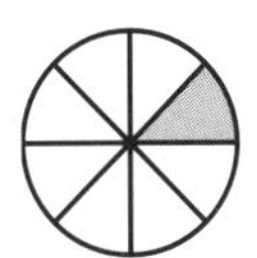

G 

H 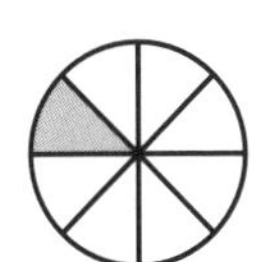

J 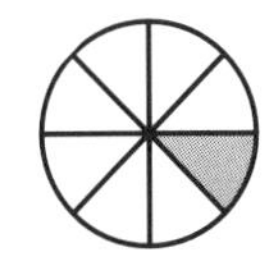

3 **What figure comes next in this pattern?**

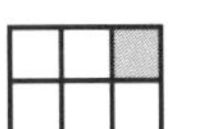
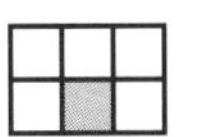
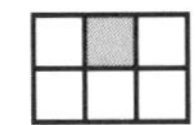

A 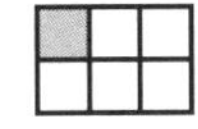

B 

C 

D 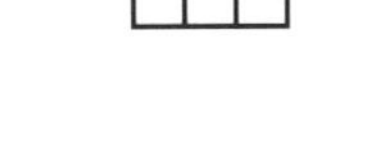

4 **The "?" shows where a figure is missing in this pattern. What figure is missing?**

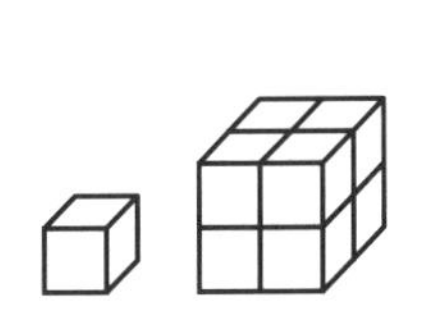
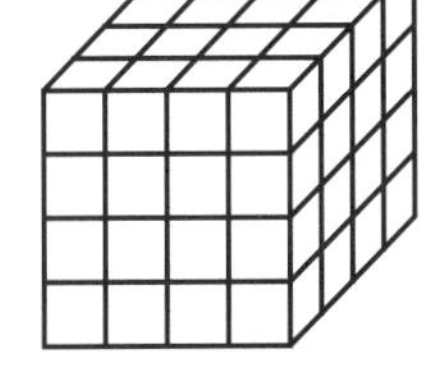

F 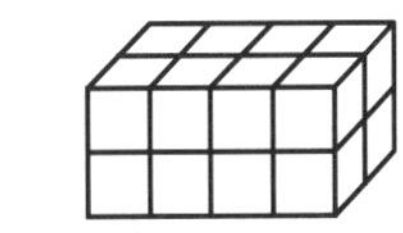

H 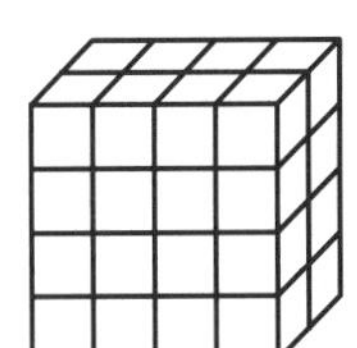

G 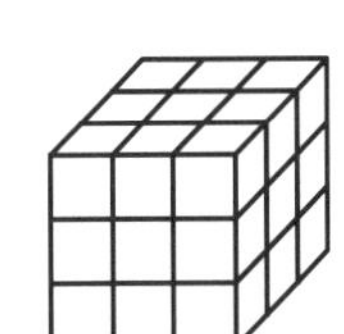

J 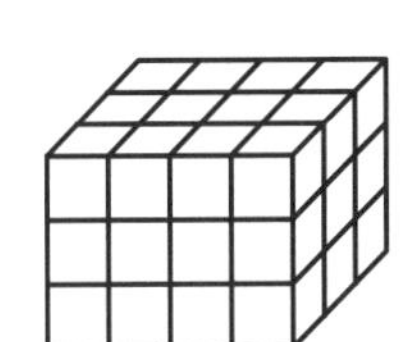

5 **How will the number of circles change for the next figure in this pattern?**

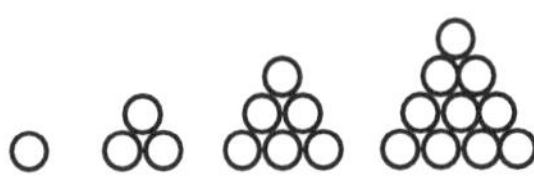

A Triple
B Double
C Increase by 5
D Increase by 6

Name ________________________________

Practice Objective 18

DIRECTIONS

Read each question and choose the best answer. Then mark the space for the answer you have chosen.

SAMPLE

Which property is used in the sentence below?

2 + (8 + 9) = (2 + 8) + 9

A Commutative Property

B Associative Property

C Distributive Property

D Identity Property

1 **Which property is used in the sentence below?**

7 + 15 = 15 + 7

A Commutative Property

B Associative Property

C Distributive Property

D Identity Property

2 **Which property is used in the sentence below?**

35 + 0 = 35

F Associative Property

G Commutative Property

H Distributive Property

J Identity Property

3 **Which property is used in the sentence below?**

189 × 1 = 189

A Commutative Property

B Associative Property

C Distributive Property

D Identity Property

4 **Which property is used in the sentence below?**

8 × 23 = 8 × (20 + 3)

F Associative Property

G Commutative Property

H Distributive Property

J Identity Property

5 **What inverse operation do you use when solving a division question?**

A Addition

B Subtraction

C Multiplication

D Division

Name ______________________________

Practice Objective 18

DIRECTIONS

Read each question and choose the best answer. Then mark the space for the answer you have chosen.

SAMPLE

Which number sentence goes with $56 - n = 16$?

A $56 + n = 16$
B $n + 16 = 56$
C $n \times 16 = 56$
D $n - 16 = 56$

1. **Which number sentence goes with $35 \div 7 = n$?**

A $7 \div n = 35$
B $35 \times 7 = n$
C $n - 7 = 35$
D $n \times 7 = 35$

2. **Which number sentence goes with $129 + 387 = n$?**

F $n + 129 = 387$
G $n - 387 = 129$
H $387 - n = 129$
J $387 \div 3 = n$

3. **Which number sentence goes with $30 \times 15 = n$?**

A $n \times 15 = 30$
B $n + 15 = 30$
C $30 \div n = 15$
D $n \div 30 = 15$

4. **If $25 \times 38 = n \times 25$, then $n =$**

F 13
G 38
H 63
J 494

5. **If $(15 + 6) + 10 = 15 + (n + 10)$, then $n =$**

A 6
B 9
C 21
D 31

6. **If $0.5 \times n = 0.5$, then $n =$**

F 1
G 0.5
H 0.25
J 0.1

7. **If $\frac{11}{15} \times n = 0$, then $n =$**

A 1
B $\frac{3}{5}$
C $\frac{4}{15}$
D 0

8. **If $6 \times 23 = n \times (20 + 3)$, then $n =$**

F 120
G 23
H 18
J 6

9. **If $4 \times 59 = 4 \times (50 + n)$, then $n =$**

A 9
B 36
C 200
D 236

10. **If $7 \times 85 = 7 \times (n + 5)$, then $n =$**

F 35
G 56
H 80
J 595

Name ______________________________

Practice Objective 19

DIRECTIONS

Read each question and choose the best answer. Then mark the space for the answer you have chosen.

SAMPLE

If $x + 5 = 8$, then

A $x - (10 - 5) = 8$

B $x - (5 + 8) = 8$

C $x + (10 \div 2) = 8$

D $x + (4 + 4) = 8$

1 If $k + 2 = 10$, then

A $k + (5 + 5) = 10$

B $k + (12 - 10) = 10$

C $k - (10 - 2) = 10$

D $k - (2 + 10) = 10$

2 If $t - 3 = 15$, then

F $t + (5 - 3) = 15$

G $t + (5 - 2) = 15$

H $t - (5 - 2) = 15$

J $t - (15 - 5) = 3$

3 If $30 = h - 10$, then

A $10 = 30 - (h + 5)$

B $10 = h + (17 + 13)$

C $30 = h - (30 + 10)$

D $30 = h + (^-5 + {}^-5)$

4 If $9 = 5 + 4$, then

F $9 \div 5 = (5 + 4) - 9$

G $6 + 9 = 6 \times 5 \times 4$

H $9 + 3 = (9 \div 9) \div 3$

J $8 \times 9 = 8 \times (5 + 4)$

5 If $8 = 10 - 2$, then

A $3 \times 8 = 3 \times (2 + 10)$

B $8 + 5 = (10 - 2) + 5$

C $8 \div 4 = (2 + 2) \div 4$

D $8 - 7 = (4 \times 3) - 7$

6 If $6 = 2 + 4$, then

F $6 + 5 = (3 \times 3) + 5$

G $9 - 6 = 9 - (2 + 4)$

H $3 \times 6 = 6 \times (3 \times 3)$

J $6 \div 3 = 6 \div (3 - 1)$

7 If $5 = 3 + 2$, then

A $3 \times 5 = 3 \times (5 - 2)$

B $5 - 4 = (3 + 2) - 4$

C $8 - 5 = 8 + (5 - 1)$

D $15 \div 5 = 15 \div (6 - 3)$

8 If $7 = 9 - 2$, then

F $7 - 4 = (9 - 7) - 4$

G $7 \times 9 = (9 - 7) \times 9$

H $5 + 7 = 5 + (7 - 5)$

J $6 \times 7 = 6 \times (9 - 2)$

Name ______________________________

DIRECTIONS

Read each question and choose the best answer. Then mark the space for the answer you have chosen.

SAMPLE

How many zeros are there in the standard form of 10^6?

A 4 zeros

B 5 zeros

C 6 zeros

D 7 zeros

1 **How many zeros are in the standard form of 10^4?**

A 3 zeros

B 4 zeros

C 5 zeros

D 6 zeros

2 **Which expression represents 5^2?**

F $5 + 2$

G 5×2

H 5×5

J $2 \times 2 \times 2 \times 2 \times 2$

3 **What is $4 \times 4 \times 4 \times 4 \times 4$ written in exponent form?**

A 4^3

B 3^4

C 4^5

D 5^4

4 **What is $6 \times 6 \times 6 \times 6$ written in exponent form?**

F 6^6

G 6^4

H 4^6

J 4^4

5 **What is the value of 3^3?**

A 9

B 12

C 27

D 81

6 **What is the value of 6^3?**

F 216

G 42

H 36

J 18

7 **The formula for the area of a square is $A = s^2$. What is the area of a square flower bed whose side measures 2.5 meters?**

A 4 square meters

B 5 square meters

C 1.25 square meters

D 6.25 square meters

Name ______________________________

Practice Objective 21

DIRECTIONS

Read each question and choose the best answer. Then mark the space for the answer you have chosen.

SAMPLE

Find $5 + 3 \times 2$.

A 10 C 14
B 11 D 16

1 **Find $3 \times (9 - 4)$.**

A 15 C 27
B 23 D 39

2 **Find $20 - (3 + 2)$.**

F 25 H 17
G 19 J 15

3 **Find $8 + 16 \div 4$.**

A 4 C 12
B 6 D 25

4 **Find $20 - 15 \div 5$.**

F 0 H 17
G 1 J 23

5 **Find $15 - 3 + 8 - 2$.**

A 18 C 6
B 16 D 2

6 **Find $20 - 4^2$.**

F 36 H 12
G 24 J 4

7 **Find $3 + 2 \times 5 - 1$**

A 11 C 20
B 12 D 24

8 **Find $6 + 18 \div 3 - 2$.**

F 6 H 10
G 8 J 24

9 **Find $8 + (10 \div 5) \times 2$.**

A 10 C 20
B 12 D 32

10 **Find $3^2 + (4 \times 2)$.**

F 26 H 14
G 17 J 8

11 **Find $3 \times 4 \times (5 - 3)^2$.**

A 60 C 48
B 57 D 36

Name ______________________________

Practice Objective 22

DIRECTIONS

Read each question and choose the best answer. Then mark the space for the answer you have chosen.

SAMPLE

On a map, which direction is to the left?

A North
B South
C East
D West

Use Shing's map below to help you answer questions 1–5. Each square equals one city block.

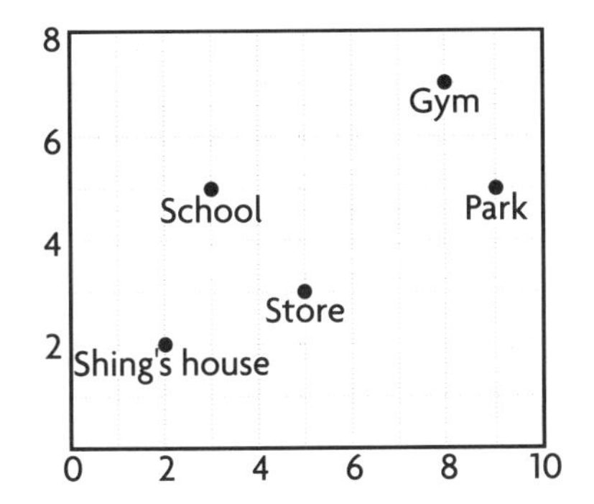

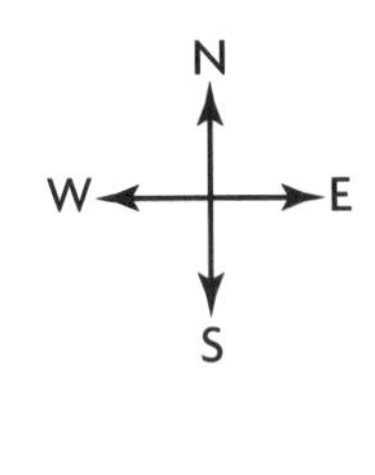

1 **Shing left his house and walked 2 blocks north, 7 blocks east, and 1 block north. Where was he?**

A Park
B Gym
C School
D Store

2 **Jorge left the gym and walked 1 block north, 5 blocks west, and 3 blocks south. Where did he go?**

F School
G Park
H Gym
J Shing's house

3 **Which location is represented by the ordered pair (9,5)?**

A Park.
B Gym
C School
D Store

4 **What ordered pair represents the location of the store?**

F (10,5)
G (5,10)
H (5,3)
J (3,5)

5 **What ordered pair represents the location of the school?**

A (6,9)
B (9.6)
C (5,3)
D (3,5)

Use the graph to answer questions 6 and 7.

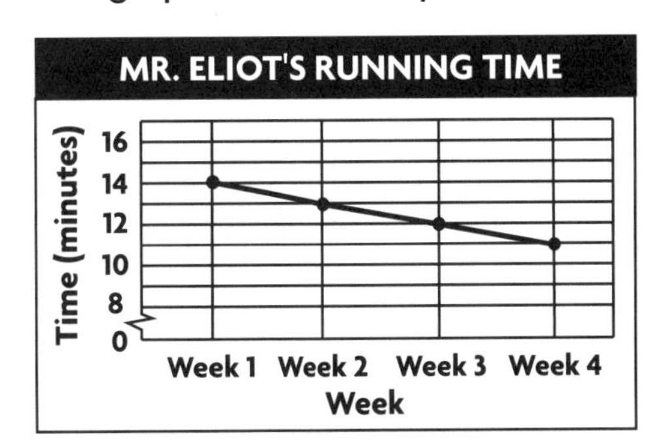

6 **During which week was Mr. Eliot's running time 12 minutes?**

F Week 1
G Week 2
H Week 3
J Week 4

7 **What was Mr. Eliot's running time for Week 1?**

A 11 minutes
B 12 minutes
C 13 minutes
D 14 minutes

Name ______________________________

DIRECTIONS

Read each question and choose the best answer. Then mark the space for the answer you have chosen.

SAMPLE

How do you locate the point $(^-4,1)$ on the coordinate grid? Begin at 0.

A Go up 4 spaces, and go right 1 space.

B Go down 4 spaces, and go right 1 space.

C Go left 4 spaces, and go down 1 space.

D Go left 4 spaces, and go up 1 space.

Use the coordinate plane to help you answer questions 1–2.

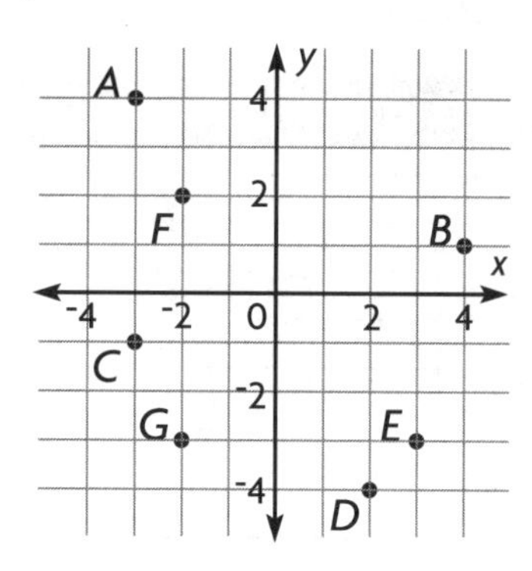

1 **Which point is located by the ordered pair $(3,^-3)$?**

A Point *D* **C** Point *F*

B Point *E* **D** Point *G*

2 **Which point is located by the ordered pair $(^-3,^-1)$**

F Point *A* **H** Point *C*

G Point *B* **J** Point *D*

Use the coordinate plane to help you answer questions 3–6.

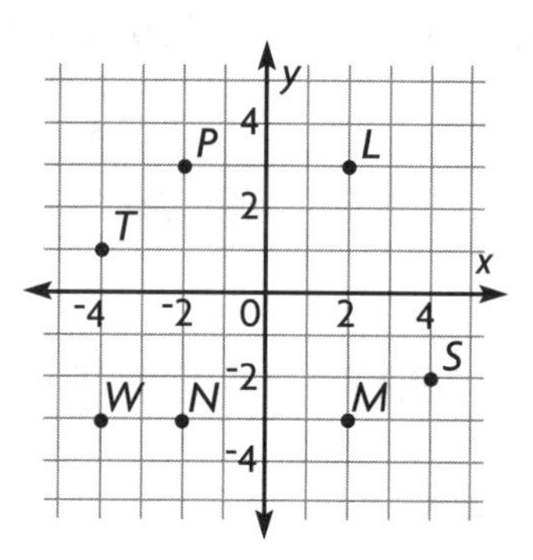

3 **Which point is located by the ordered pair $(^-4,1)$?**

A Point *Q* **C** Point *S*

B Point *R* **D** Point *T*

4 **Which point is located by the ordered pair (2,3)?**

F Point *L* **H** Point *N*

G Point *M* **J** Point *P*

5 **What is the ordered pair for point *M*?**

A $(2,^-3)$ **C** $(^-2,^-3)$

B $(^-2,3)$ **D** (2,3)

6 **What is the ordered pair for point *P*?**

F $(2,^-3)$ **H** (2,3)

G $(^-2,^-3)$ **J** $(^-2,3)$

Name ______________________________

Practice Objective 23

DIRECTIONS

Read each question and choose the best answer. Then mark the space for the answer you have chosen.

SAMPLE

A number machine uses the rule "Multiply by 8" to change numbers into other numbers. What is the output?

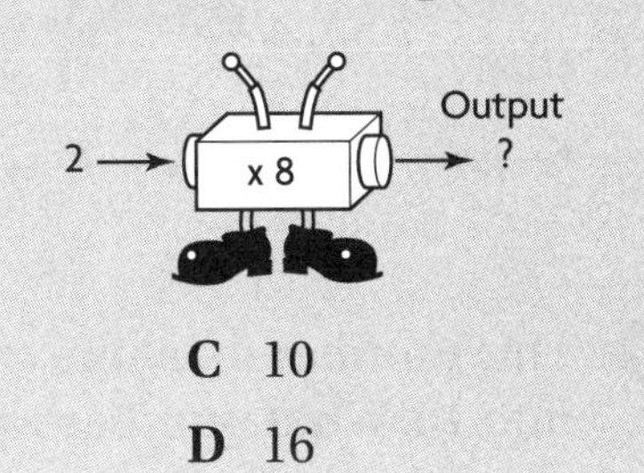

A 4
B 6
C 10
D 16

1 **A number machine uses the rule "Add 5" to change numbers into other numbers. What is the output?**

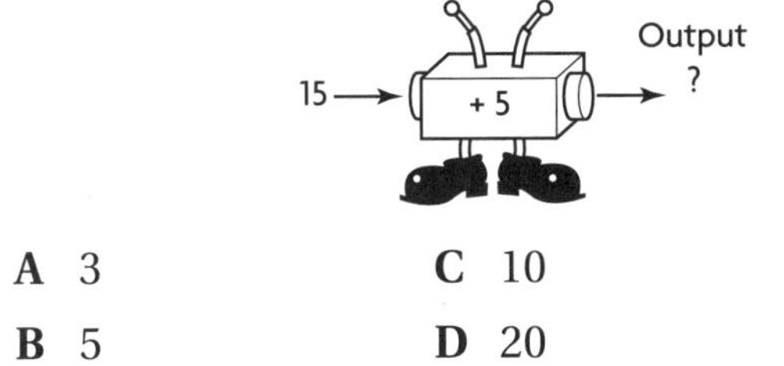

A 3
B 5
C 10
D 20

2 **A number machine uses the rule "Subtract 12" to change numbers into other numbers. What is the output?**

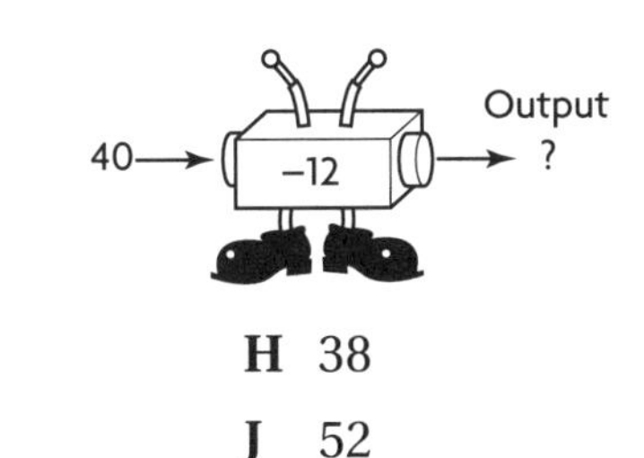

F 4
G 28
H 38
J 52

3 **A number machine uses the rule "Divide by 4" to change numbers into other numbers. What is the output?**

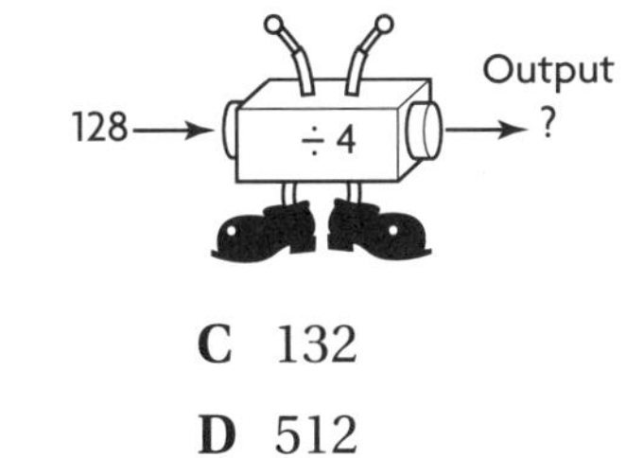

A 32
B 124
C 132
D 512

4 **What is the output?**

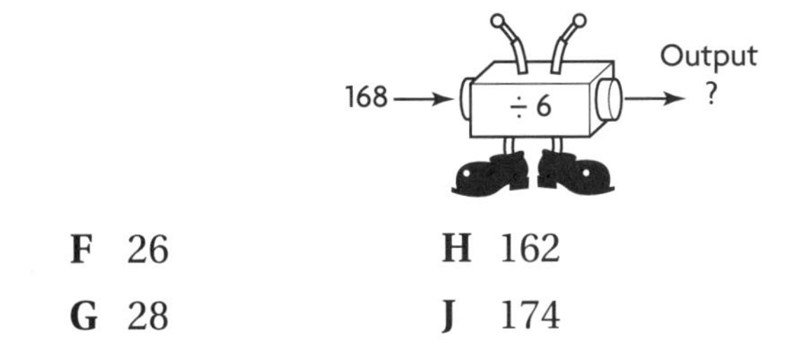

F 26
G 28
H 162
J 174

5 **What is the output?**

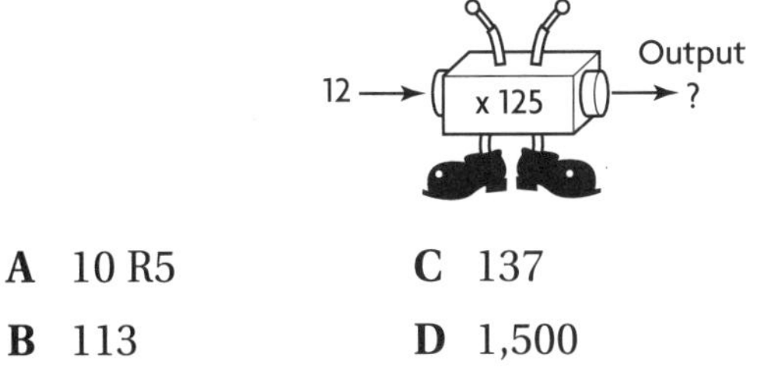

A 10 R5
B 113
C 137
D 1,500

6 **What is the output?**

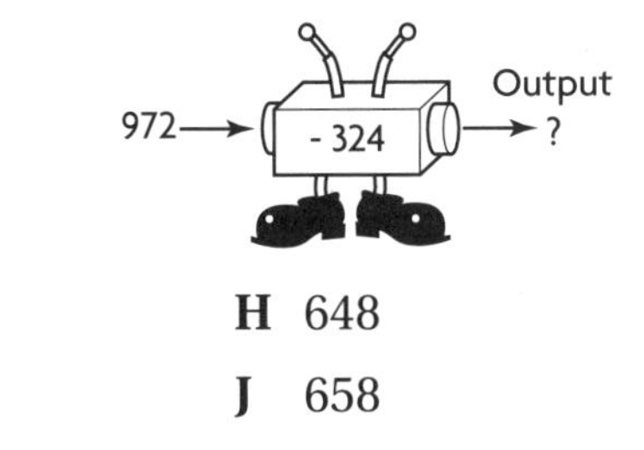

F 3
G 558
H 648
J 658

Name ______________________

Practice Objective 23

DIRECTIONS

Read each question and choose the best answer. Then mark the space for the answer you have chosen.

SAMPLE

A number machine uses a secret rule to change numbers into other numbers.

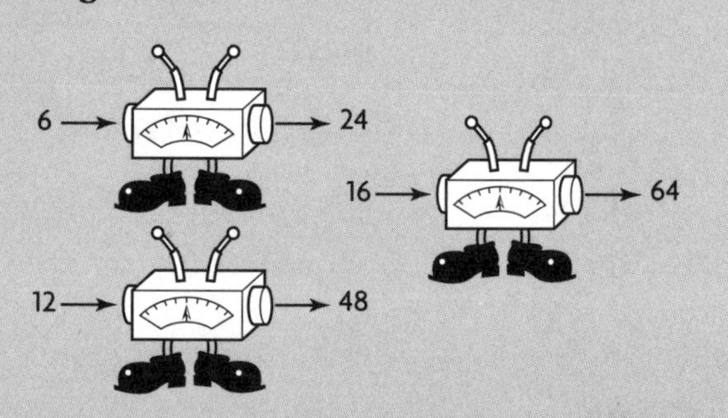

This number machine changed 6 into 24, 12 into 48, and 16 into 64. What number would 8 be changed into?

A 24
B 32
C 48
D 56

1 **A number machine uses a secret rule to change numbers into other numbers.**

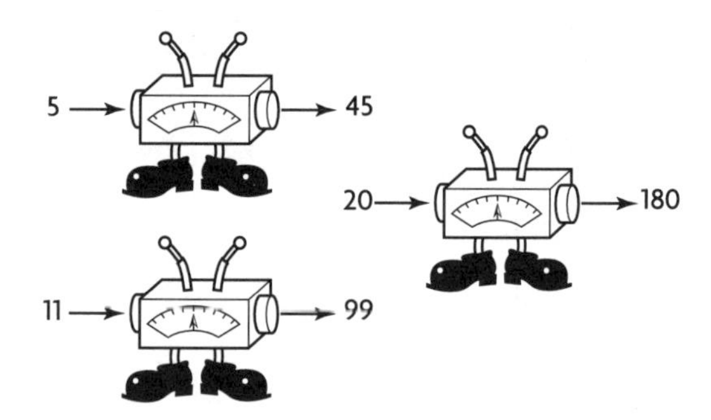

This number machine changed 5 into 45, 11 into 99, and 20 into 180. What number would 7 be changed into?

A 140
B 77
C 63
D 42

2 **This number machine changed 9 into 2. What number would 15 be changed into?**

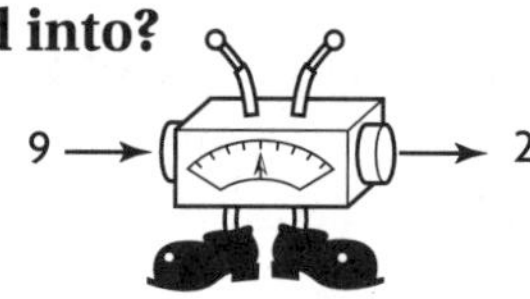

F 8
G 24
H 32
J 135

3 **This number machine changed 100 into 10. What number would 25 be changed into?**

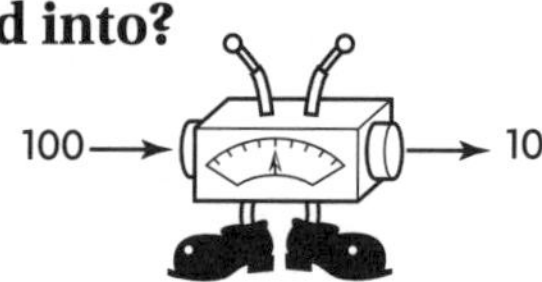

A 0.25
B 2.05
C 2.5
D 250

4 **This number machine changed 8 into 20. What number would 4 be changed into?**

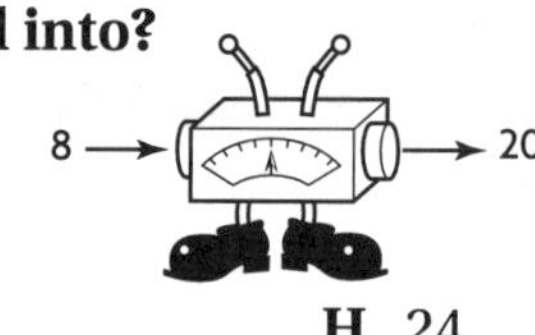

F 9
G 10
H 24
J 29

5 **This number machine changed $6\frac{1}{4}$ into $4\frac{1}{2}$. What number would $3\frac{1}{4}$ be changed into?**

A $1\frac{1}{2}$
B $2\frac{1}{4}$
C 4
D $4\frac{1}{4}$

Name ______________________________

DIRECTIONS

Read each question and choose the best answer. Then mark the space for the answer you have chosen.

SAMPLE

The angle formed by the alligator's mouth is __?__.

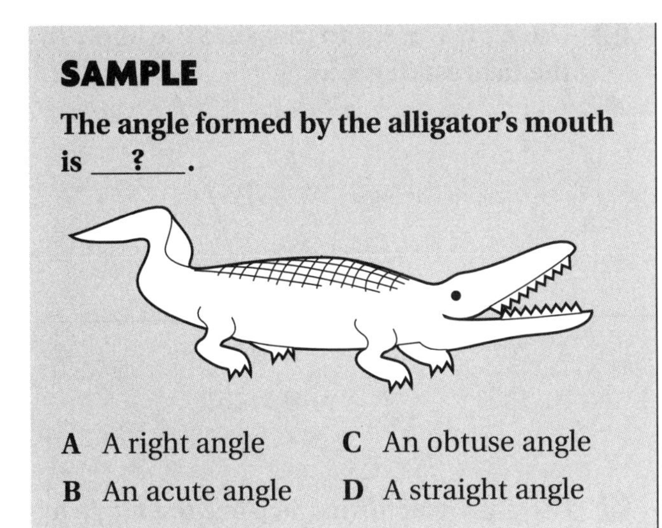

A A right angle
B An acute angle
C An obtuse angle
D A straight angle

1 The angle formed by the blades of the scissors is __?__.

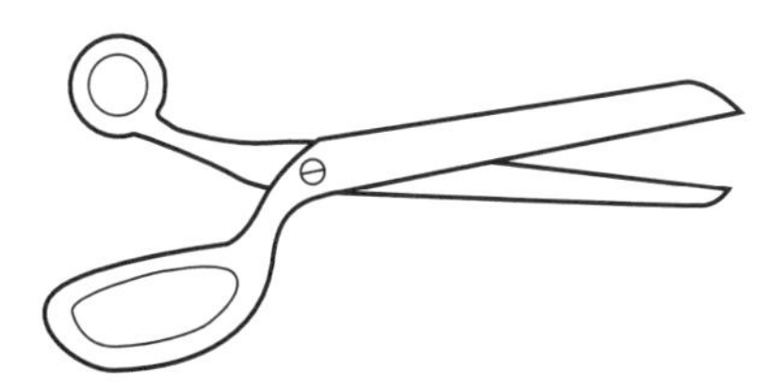

A A right angle
B An acute angle
C An obtuse angle
D A straight angle

2 What kind of angle is formed by the two logs?

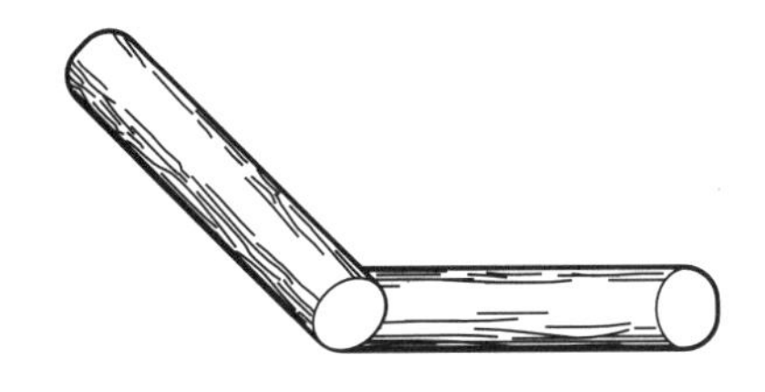

F Right angle
G Acute angle
H Obtuse angle
J Straight angle

3 What kind of angle is formed by two stair steps?

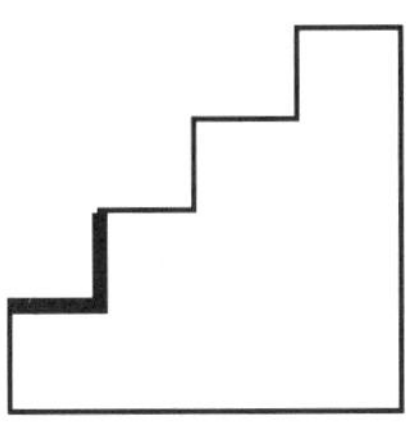

A Right angle
B Acute angle
C Obtuse angle
D Straight angle

4 What kind of angle is formed where the bases of the picture frames meet?

F Right angle
G Acute angle
H Obtuse angle
J Straight angle

5 At which time do the hands of a clock form a straight angle?

A 9:00
B 6:30
C 6:00
D 3:00

6 Barbara is making a scale drawing of her home. What kind of angle will she use when drawing the square corners of the rooms?

F Right angle
G Acute angle
H Obtuse angle
J Straight angle

Name ____________________

Practice Objective 24

DIRECTIONS

Read each question and choose the best answer. Then mark the space for the answer you have chosen.

Use a protractor to measure the angle to the nearest degree.

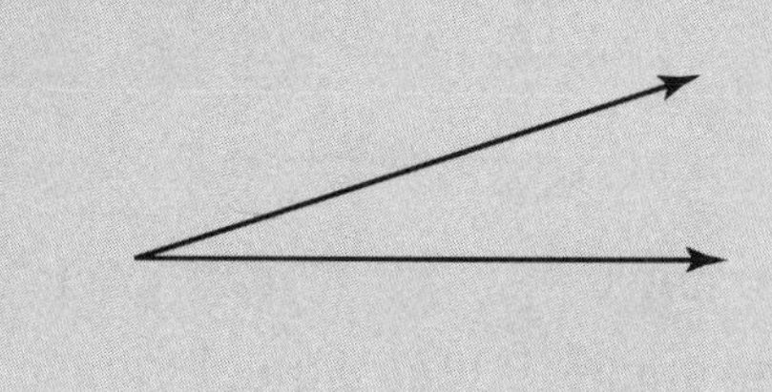

A 18° C 135°
B 35° D 145°

1 **Use a protractor to measure the angle to the nearest degree.**

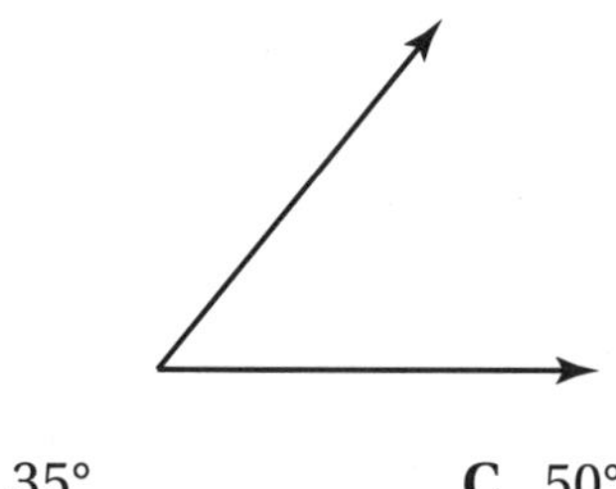

A 35° C 50°
B 45° D 55°

2 **Use a protractor to measure the angle to the nearest degree.**

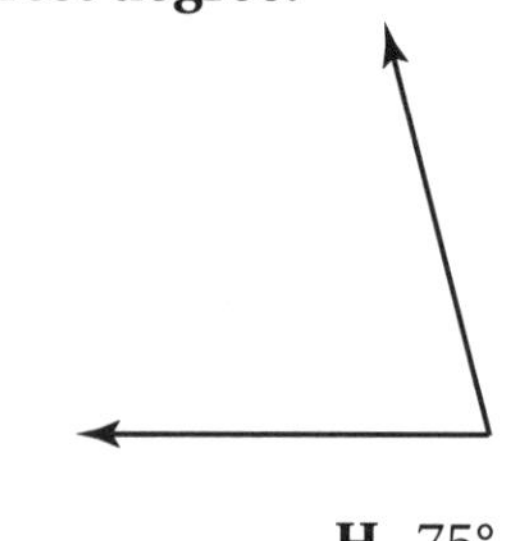

F 65° H 75°
G 70° J 105°

3 **Use a protractor to measure the angle to the nearest degree.**

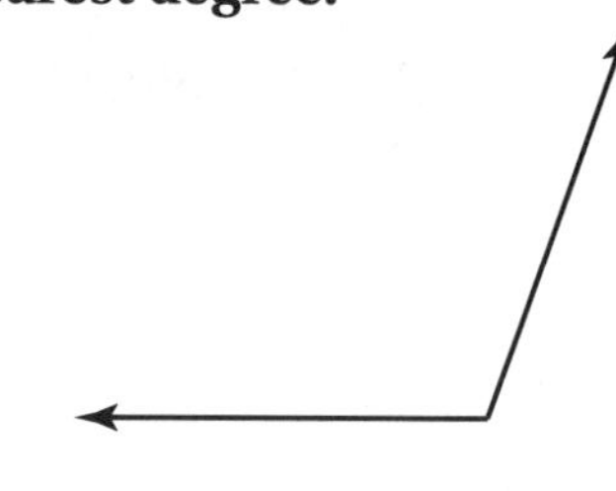

A 70° C 110°
B 100° D 115°

4 **Use a protractor to measure the angle to the nearest degree.**

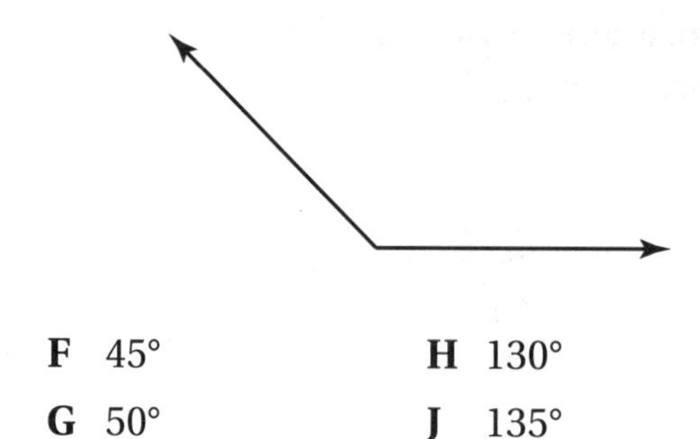

F 45° H 130°
G 50° J 135°

5 **Use a protractor to measure the angle to the nearest degree.**

A 35° C 100°
B 90° D 180°

Name ______________________

DIRECTIONS

Read each question and choose the best answer. Then mark the space for the answer you have chosen.

Use the figure below for the sample question and question 1.

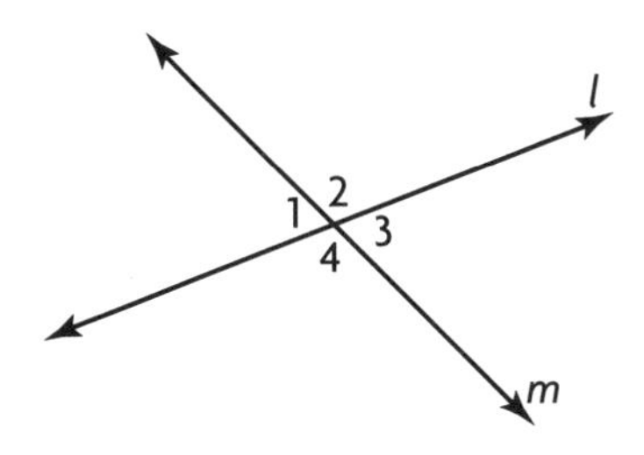

SAMPLE

Which pair of angles are adjacent?

A Angle 1 and Angle 3
B Angle 2 and Angle 4
C Angle 2 and Angle 3
D Line *l* and line *m*

1 **Which pair of angles are vertical angles?**

A Angle 1 and Angle 3
B Angle 1 and Angle 2
C Angle 2 and Angle 3
D Angle 3 and Angle 4

Lines *l* and *m* intersect. Use the figure below to help you answer questions 2–4.

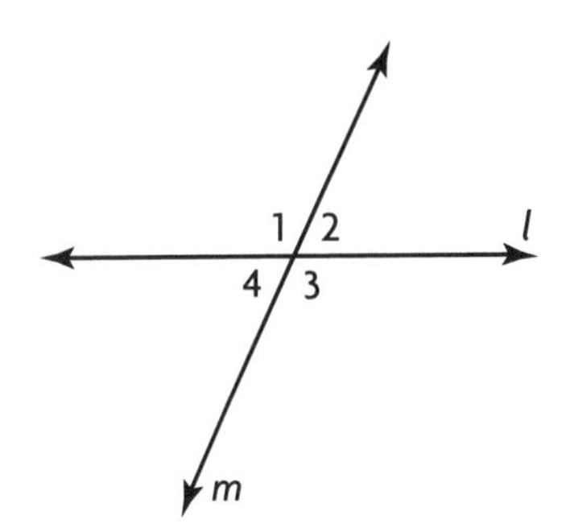

2 **Which of these angles and Angle 4 form a pair of adjacent angles?**

F Angle 1 only
G Angle 2 only
H Either angle 2 or angle 4
J Either angle 1 or angle 3

3 **Which of these angles and Angle 1 form a pair of vertical angles?**

A Angle 1
B Angle 2
C Angle 3
D Angle 4

4 **Which term describes Angle 2 and Angle 4?**

F Adjacent angles
G Vertical angles
H Right angles
J Obtuse angles

Name ______________________________

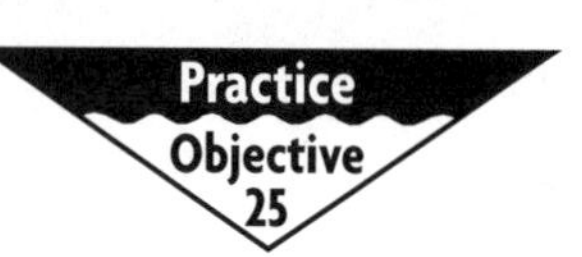

DIRECTIONS

Read each question and choose the best answer. Then mark the space for the answer you have chosen.

SAMPLE

__?__ are opposite angles formed when two lines intersect. They are always congruent.

A Congruent line segments

B Perpendicular line segments

C Adjacent angles

D Vertical angles

1 **__?__ have the same vertex and a common side.**

A Adjacent angles

B Vertical angles

C Right angles

D Congruent line segments

2 **In the diagram ∠1 and ∠3 are called __?__.**

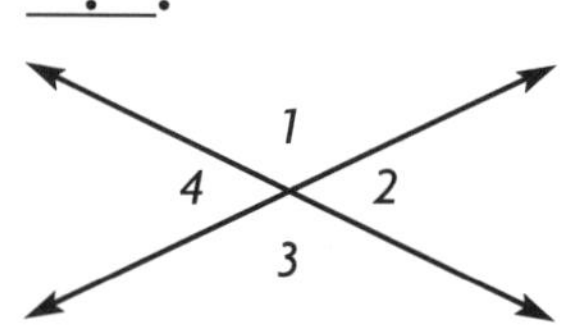

F Acute angles

G Right angles

H Adjacent angles

J Vertical angles

3 **In the diagram, ∠2 and ∠3 are called __?__.**

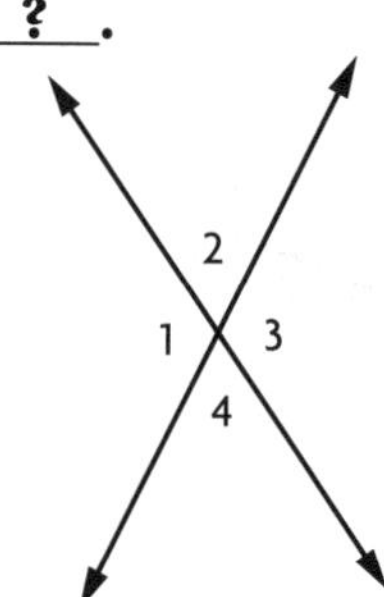

A Congruent angles

B Opposite angles

C Adjacent angles

D Vertical angles

Use the diagram below to help you answer questions 4 and 5.

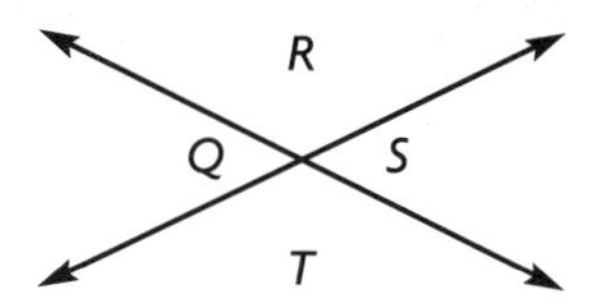

4 **In the diagram, ∠T and ∠Q are __?__.**

F Adjacent

G Vertical

H Congruent

J Opposite

5 **Which of the following choices would not complete the following sentence correctly.**

In the diagram, ∠S and ∠Q are __?__.

A Adjacent

B Vertical

C Congruent

D Opposite

Name ______________________________

DIRECTIONS

Read each question and choose the best answer. Then mark the space for the answer you have chosen.

SAMPLE

Which triangle below is a regular polygon?

A Equilateral triangle
B Right triangle
C Acute triangle
D Obtuse triangle

1 **What is the name for this polygon?**

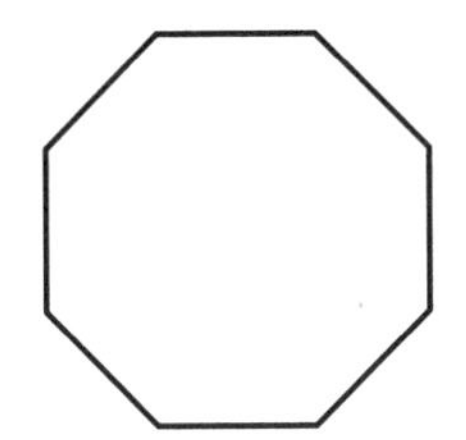

A Pentagon
B Trapezoid
C Hexagon
D Octagon

2 **How many sides does a pentagon have?**

F 5 **H** 7
G 6 **J** 8

For questions 3–4, use the figures below.

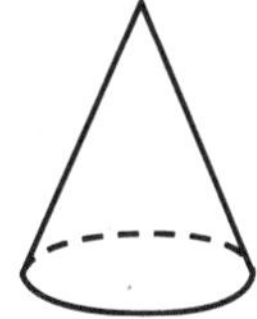

Figure A

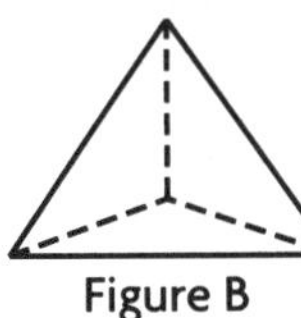
Figure B

Figure C

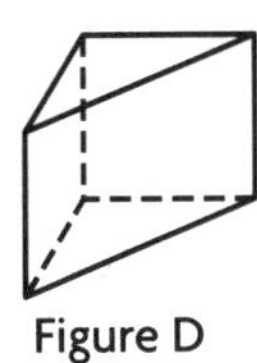
Figure D

3 **Which figure is a cone?**

A Figure A **C** Figure C
B Figure B **D** Figure D

4 **Which figure is a triangular prism?**

F Figure A **H** Figure C
G Figure B **J** Figure D

For questions 5–6, use the figure below.

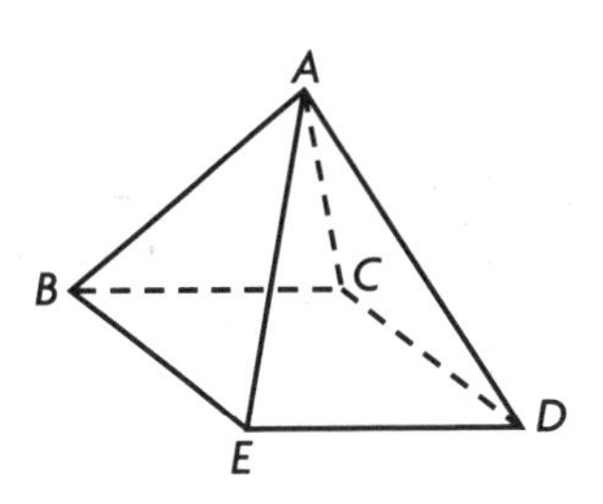

5 **How many faces does this solid figure have?**

A 4 **C** 6
B 5 **D** 7

6 **How many edges does this figure have?**

F 5 **H** 7
G 6 **J** 8

7 **The top, bottom, and side views of a solid figure are all squares. What kind of figure is it?**

A Hexagonal prism
B Cylinder
C Triangular pyramid
D Cube

Name ______________________________

Practice Objective 27

DIRECTIONS

Read each question and choose the best answer. Then mark the space for the answer you have chosen.

SAMPLE

In which figure is the dashed line a line of symmetry?

A 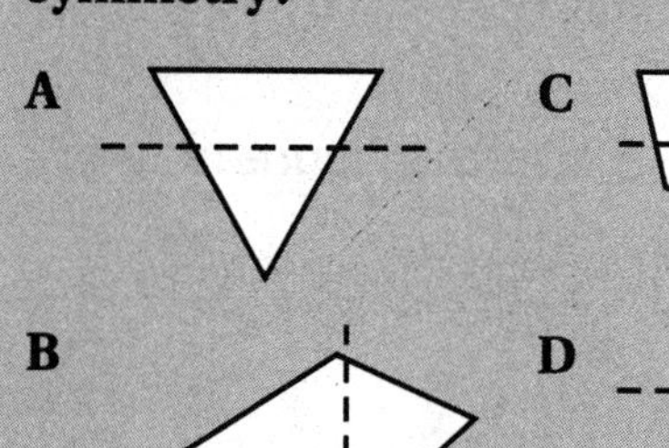

C 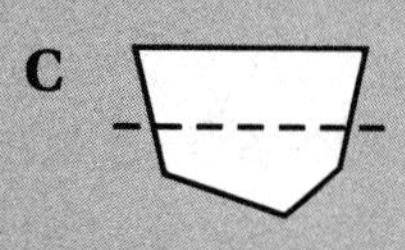

B D

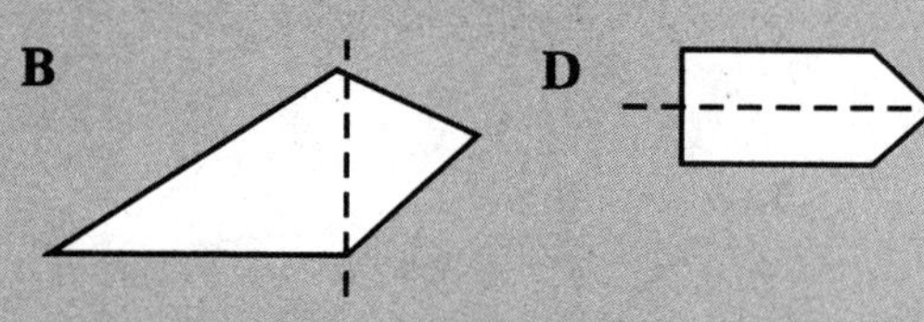

1 **In which figure is the dashed line a line of symmetry?**

A 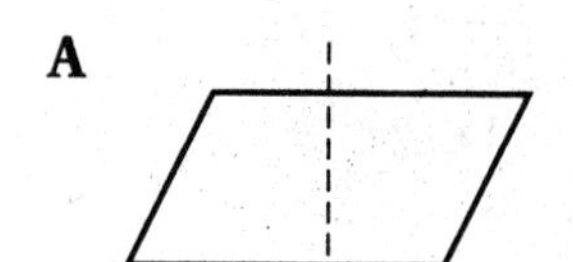

C 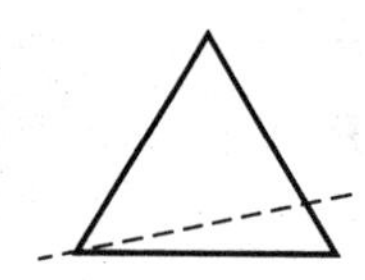

B 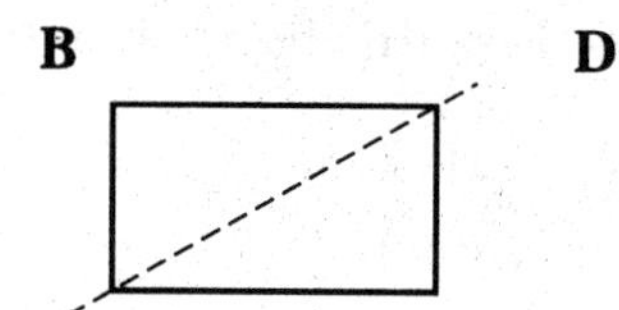

D 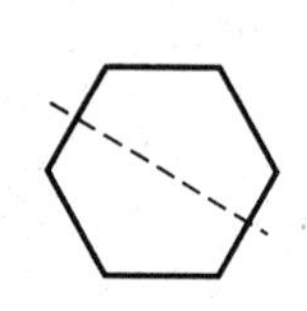

2 **How many lines of symmetry does this figure have?**

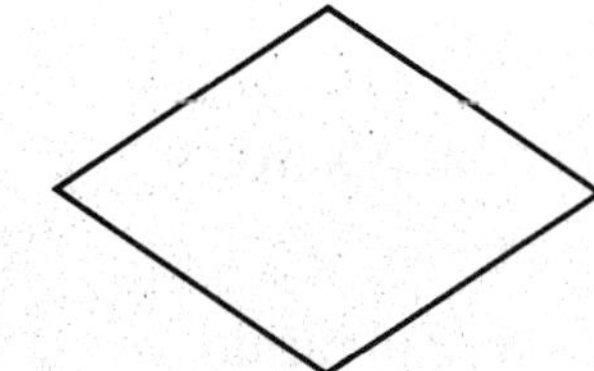

F 0 lines

G 1 line

H 2 lines

J 3 lines

3 **How many lines of symmetry does this figure have?**

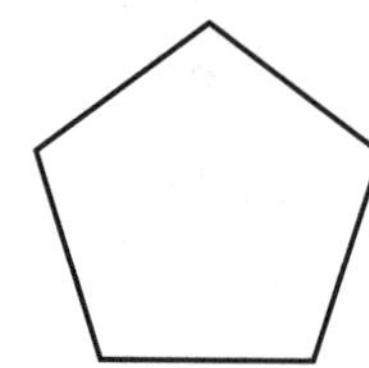

A 6 lines

B 5 lines

C 4 line

D 1 line

4 **How many lines of symmetry does this figure have?**

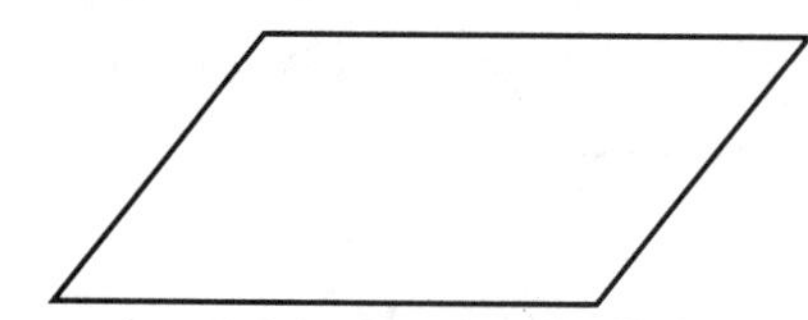

F 0 lines

G 1 line

H 2 lines

J 5 lines

5 **Toi bought a brownie that was in the shape of a rectangle. She cut the brownie into pieces along its lines of symmetry. How many pieces did she have then?**

A 1 piece

B 2 pieces

C 4 pieces

D 8 pieces

Name ______________________________

DIRECTIONS

Read each question and choose the best answer. Then mark the space for the answer you have chosen.

SAMPLE

Which partial drawing could be completed as a square by drawing exactly one line segment?

A

C

B

D

1 **Which partial drawing could be completed as a rectangle by drawing exactly one line segment?**

A

C

B

D

2 **Which partial drawing could be completed as a parallelogram by drawing one line segment?**

F

H

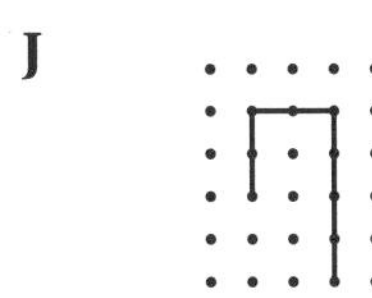

G

J

3 **After drawing the angle below, what kind of triangle could you form by drawing one line segment?**

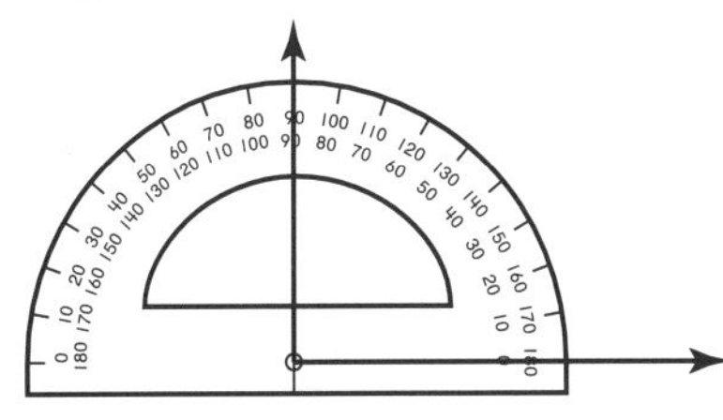

A Acute triangle

B Equilateral triangle

C Obtuse triangle

D Right triangle

4 **Two 60° angles were measured on the line segment below. What kind of triangle could you form if you extend the segments until they intersect?**

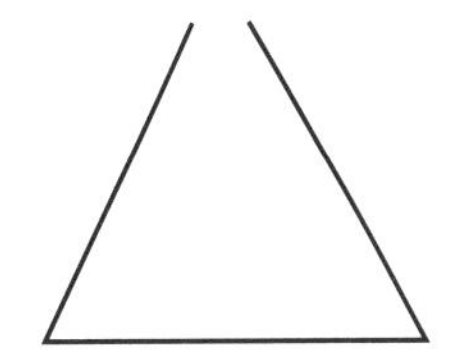

F Right triangle

G Scalene triangle

H Obtuse triangle

J Equilateral triangle

Name ______________________________

DIRECTIONS

Read each question and choose the best answer. Then mark the space for the answer you have chosen.

SAMPLE

What construction do the figures below represent?

A Congruent line segments
B Bisected line segments
C Congruent angles
D Perpendicular line segments

1 **What construction do the figures below represent?**

A Perpendicular line segments
B Congruent angles
C Bisected line segments
D Congruent line segments

2 **What construction do the figures below represent?**

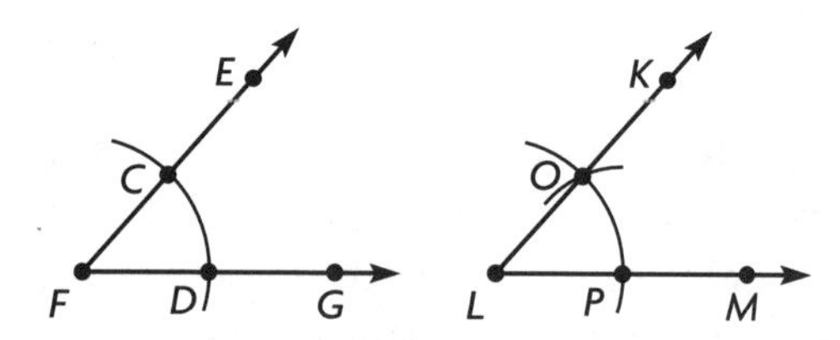

F Bisected line segments
G Parallel line segments
H Congruent angles
J Congruent line segments

3 **What construction do the figures below represent?**

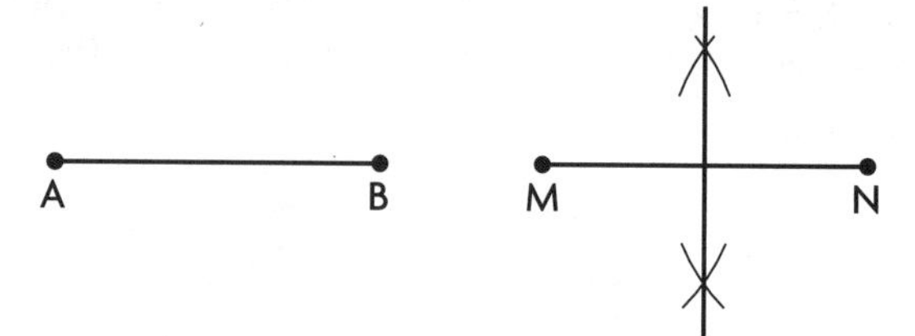

A Perpendicular line segments
B Parallel line segments
C Congruent angles
D Congruent polygons

4 **What construction do the figures below represent?**

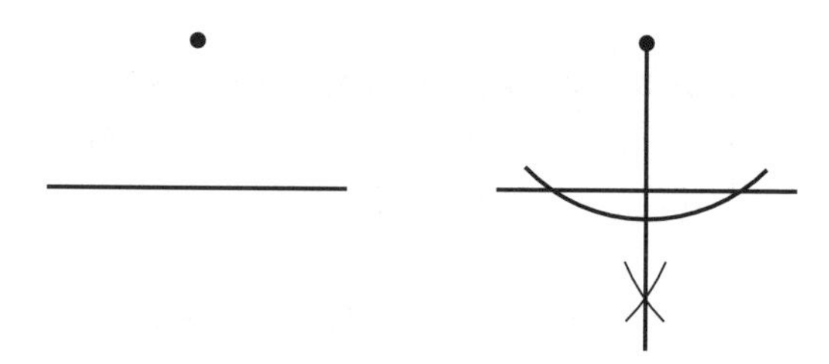

F Congruent angles
G Congruent polygons
H Parallel line segments
J Perpendicular line segments

Name ______________________________

DIRECTIONS

Read each question and choose the best answer. Then mark the space for the answer you have chosen.

SAMPLE

Which white figure shows where a gray figure would be if the paper were folded along the heavy dark line?

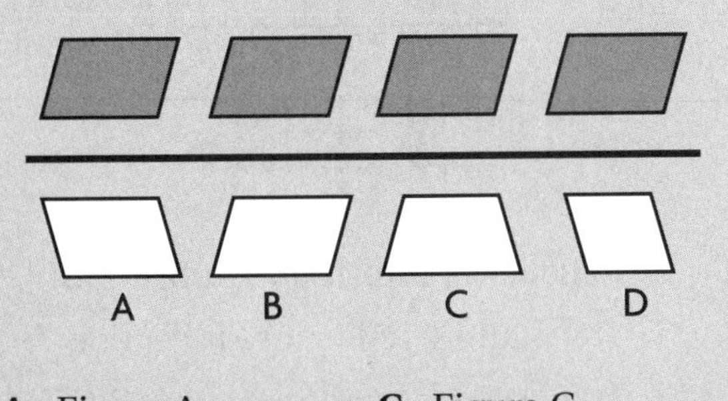

A Figure A
B Figure B
C Figure C
D Figure D

1 **Which white figure shows where the gray figure would be if the paper were folded along the heavy dark line?**

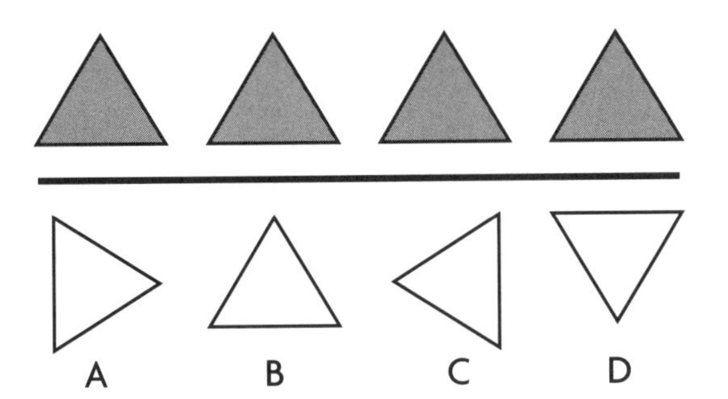

A Figure A
B Figure B
C Figure C
D Figure D

For items 2–4, use triangle *RST*. The coordinates are (1,1), (3,4), and (3,1).

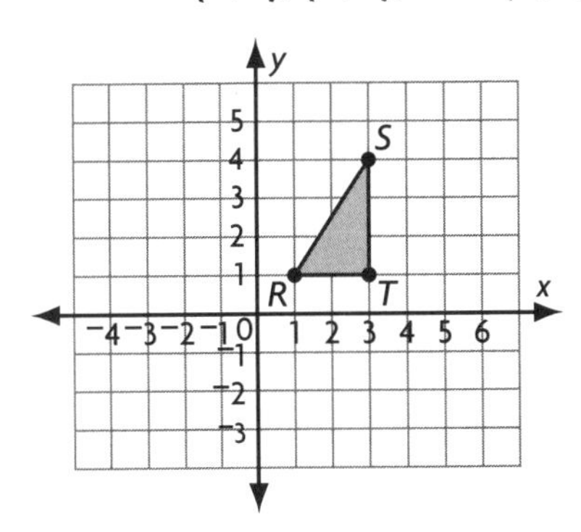

2 **If you reflect triangle *RST* across the *x*-axis, which coordinates would change?**

F Both *x*- and *y*-coordinates
G Only *x*-coordinates
H Only *y*-coordinates
J Neither *x*- nor *y*-coordinates

3 **If you reflect triangle *RST* across the *x*-axis, what would the coordinates of the new triangle be?**

A $R'(1,^-1)$, $S'(3,^-3)$ $T'(^-3,3)$
B $R'(1,^-1)$, $S'(3,-4)$, $T'(3,^-1)$
C $R'(1,1)$, $S'(^-3,1)$, $T'(^-3,4)$
D $R'(^-1,^-1)$, $S'(^-3,-4)$, $T'(^-3,^-1)$

4 **If you reflect triangle *RST* across the *y*-axis, what would the coordinates of the new triangle be?**

F $R'''(1,^-1)$, $S'''(1,4)$, $T'''(3,^-1)$
G $R'''(1,^-1)$, $S'''(3,4)$, $T'''(3,^-1)$
H $R'''(^-1,1)$, $S'''(^-1,4)$, $T'''(^-3,1)$
J $R'''(^-1,1)$, $S'''(^-3,4)$, $T'''(^-3,1)$

Name ____________________

DIRECTIONS

Read each question and choose the best answer. Then mark the space for the answer you have chosen.

SAMPLE

What moves were made to transform figure 1 into each new position?

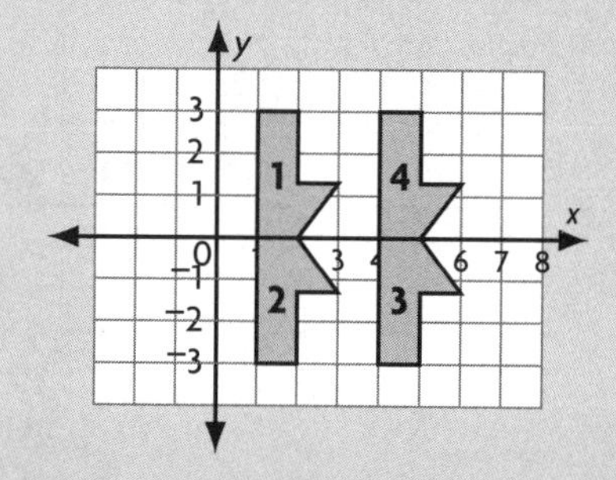

A Rotation, reflection, rotation

B Translation, reflection, translation

C Reflection, translation, reflection

D Rotation, translation, reflection

1 **What moves were made to transform figure 1 into each new position?**

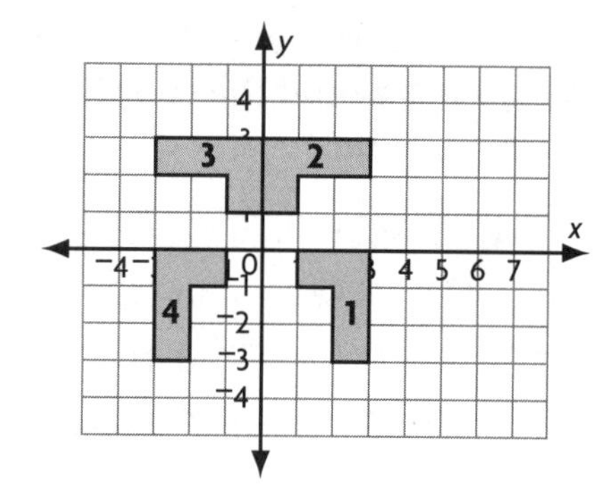

A Rotation, reflection, rotation

B Reflection, translation, reflection

C Translation, reflection, translation

D Rotation, translation, reflection

2 **What moves were made to transform figure 1 into each new position?**

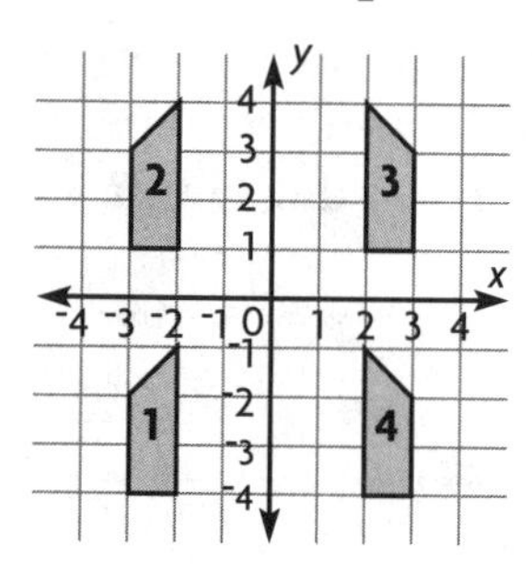

F Reflection, translation, reflection

G Translation, reflection, reflection

H Translation, rotation, translation

J Translation, reflection, translation

3 **What moves were made to transform figure 1 into each new position?**

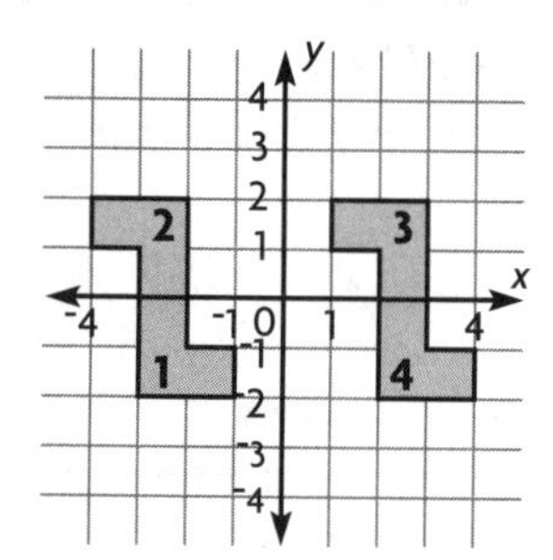

A Rotation, translation, rotation

B Translation, reflection, translation

C Rotation, reflection, rotation

D Reflection, translation, reflection

Name ______________________________

DIRECTIONS

Read each question and choose the best answer. Then mark the space for the answer you have chosen.

SAMPLE

If *C* is the center of the circle, which line segment is a radius of the circle?

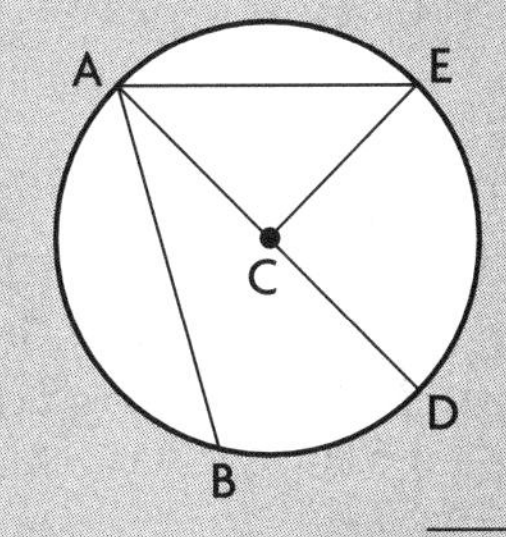

A $\overline{AB}$ C $\overline{AD}$

B $\overline{CA}$ D $\overline{AE}$

1 **If *C* is the center of the circle, which line segment is a radius of the circle?**

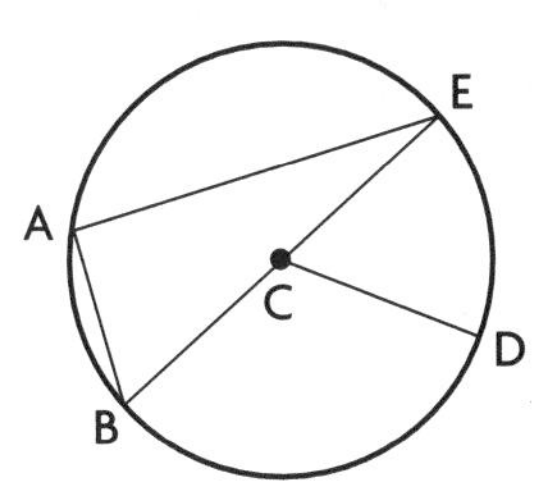

A $\overline{AB}$ C $\overline{BE}$

B $\overline{AE}$ D $\overline{CD}$

2 **If *C* is the center of the circle, which line segment is a diameter of the circle?**

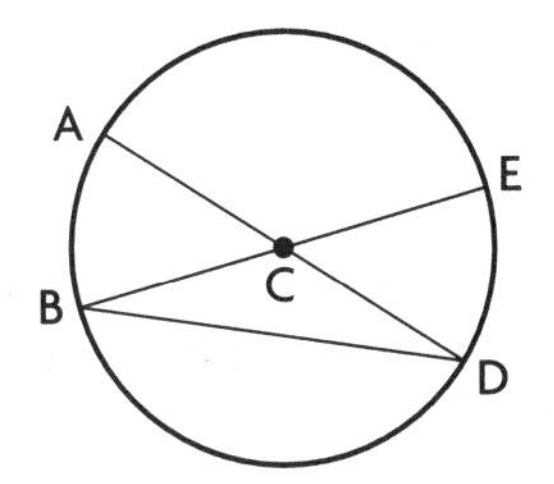

F $\overline{CA}$ H $\overline{BD}$

G $\overline{CB}$ J $\overline{BE}$

3 **If *C* is the center of the circle, which line segment is a diameter of the circle?**

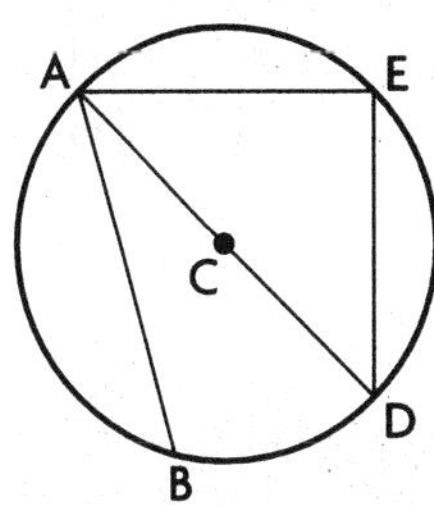

A $\overline{AB}$ C $\overline{AE}$

B $\overline{AD}$ D $\overline{DE}$

4 **If *C* is the center of the circle, which line segment is a chord of the circle?**

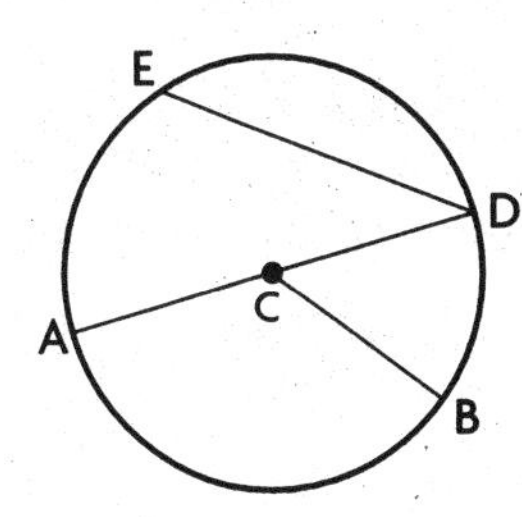

F $\overline{CD}$ H $\overline{CA}$

G $\overline{CB}$ J $\overline{ED}$

5 **The diameter of a pie measures 10 inches. If the pie is cut into six congruent pieces, what is the length of each cut side of one piece of pie?**

A 2 inches

B 5 inches

C 10 inches

D 15 inches

Name ____________________

Practice Objective 32

DIRECTIONS

Read each question and choose the best answer. Then mark the space for the answer you have chosen.

SAMPLE

Use your centimeter ruler to measure the length of the path from *A* to *B*.

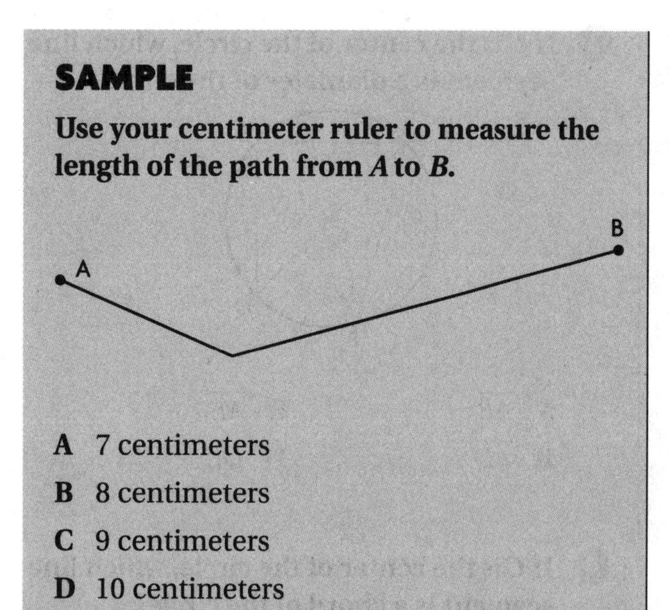

A 7 centimeters

B 8 centimeters

C 9 centimeters

D 10 centimeters

1 **Use your centimeter ruler to measure the perimeter of the figure below.**

A 2 cm **C** 15 cm

B 8 cm **D** 16 cm

2 **What is the area of the figure in item 1?**

F 8 cm^2 **H** 16 cm^2

G 15 cm^2 **J** 32 cm^2

3 **Miss Que has to go from the school to the grocery store and the library before going to a music lesson. How much difference is there between the routes *SGLM* and *SLGM*?**

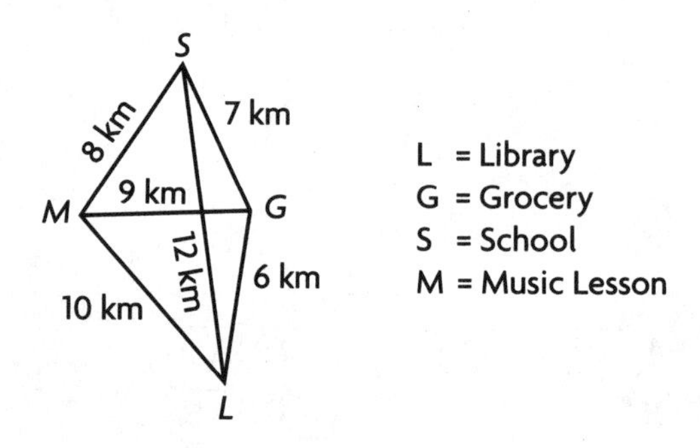

A SLGM is 1 km shorter.

B SLGM is 2 km shorter.

C SGLM is 4 km shorter.

D SGLM is 1 km shorter.

4 **If Miss Que chooses the shorter route in item 3, how many kilometers will she travel?**

F 21 km **H** 25 km

G 23 km **J** 27 km

5 **If Miss Que chooses the longer route in item 3, how many kilometers will she travel?**

A 21 km **C** 25 km

B 23 km **D** 27 km

Name ______________________

DIRECTIONS

Read each question and choose the best answer. Then mark the space for the answer you have chosen.

SAMPLE

A passenger jet travels 550 miles per hour. How far would the plane travel in 3.5 hours?

A 1,500 miles
B 1,550 miles
C 1,650 miles
D 1,925 miles

1 **A high-speed train travels 300 kilometers per hour. How far would the train travel in 2.5 hours?**

A 750 km
B 600 km
C 450 km
D 120 km

2 **Columbus is 90 miles from Louisville. How long will it take a car traveling 60 miles per hour to go from Columbus to Louisville?**

F $\frac{1}{2}$ hour
G $\frac{2}{3}$ hour
H $1\frac{1}{2}$ hours
J 3 hours

3 **Firewood is sold by the cord. A cord is a stack of wood that is 8 feet long by 4 feet wide by 4 feet high. What is the volume of a cord of wood?**

A 32 ft^3
B 128 ft^3
C 64 ft^3
D 320 ft^3

4 **If the length of each side of the cube doubles, how does its volume change?**

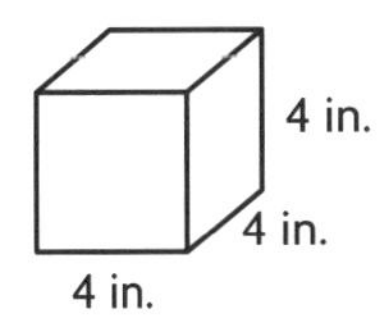

F It is twice the original cube.
G It is 4 times the original cube.
H It is 8 times the original cube.
J It is 16 times the original cube.

5 **How can you change this box so that it will hold twice as much?**

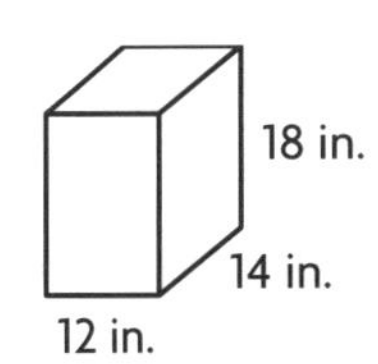

A Double only one dimension.
B Double only two dimensions.
C Double all three dimensions.
D Add 2 in. to each dimension.

6 **Emily is planning to paint just the sides of a large block that is 13 cm long, 11 cm wide, and 9 cm high. How much surface area will she paint?**

F 143 cm^2
G 432 cm^2
H 484 cm^2
J 620 cm^2

Name ______________________________

Practice Objective 33

DIRECTIONS

Read each question and choose the most appropriate measure. Then mark the space for the answer you have chosen.

SAMPLE

The length of a pen

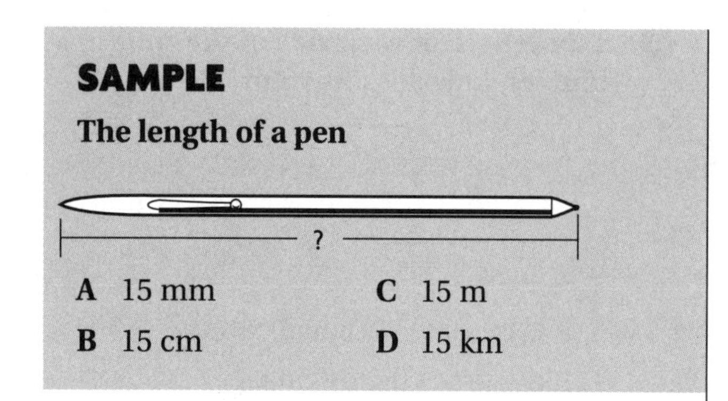

A 15 mm C 15 m

B 15 cm D 15 km

1 The length of a spoon

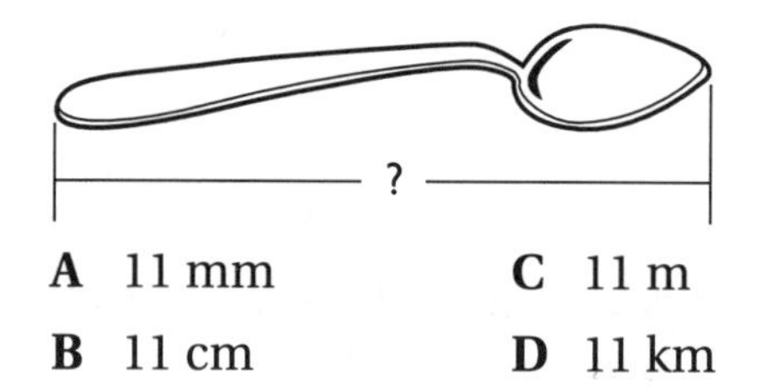

A 11 mm C 11 m

B 11 cm D 11 km

2 The distance between two airports

F 35 km H 35 cm

G 35 m J 35 mm

3 The length of a bike trail

A 100 m C 1,000 cm

B 10 km D 20 m

4 The length of a crayon

F 8 mm H 8 m

G 8 cm J 8 km

5 The diameter of a nickel

A 2 cm C 2 m

B 2 mm D 2 km

6 The distance across a bridge

F 125 cm H 2 km

G 200 mm J 200 km

7 The length of a paper clip

A 3 mm C 3 m

B 3 cm D 3 km

8 The width of your math book

F 22 km H 22 cm

G 22 m J 22 mm

9 Each of four friends measured the length of Beth's house. Which of their measurements was more precise?

A 62 m C 610 dm

B 60 m D 6,159 cm

Name ___________________________

DIRECTIONS

Read each question and choose the most appropriate measure. Then mark the space for the answer you have chosen.

SAMPLE

The capacity of a juice glass

A 100 mL **C** 10 L
B 10 mL **D** 125 L

1 **The capacity of a pitcher of punch**

A 2 mL **C** 2 L
B 20 mL **D** 20 L

2 **The capacity of a paint can**

F 4 mL **H** 4 L
G 400 mL **J** 40 L

3 **The capacity of a drinking glass**

A 300 L **C** 3 L
B 300 mL **D** 30 mL

4 **The capacity of an auto gas tank**

F 5 L **H** 50 L
G 50 mL **J** 500 L

5 **The mass of a bowling ball**

A 8 kg **C** 800 g
B 8 g **D** 80 mg

6 **The mass of an egg**

F 500 mg **H** 50 g
G 5 mg **J** 5 kg

7 **The mass of a jar of jam**

A 90 kg **C** 900 mg
B 900 g **D** 9 kg

8 **The mass of a paper clip**

F 20 kg **H** 2 g
G 2 kg **J** 20 mg

9 **The mass of stick of gum**

A 4 mg **C** 4 kg
B 4 g **D** 40 kg

10 **Each of four basketball players measured the mass of a basketball. Which of their measurements was more precise?**

F 0.6 kg **H** 0.55 kg
G 0.5 kg **J** 566 g

11 **Each of four cooks measured the capacity of a pitcher. Which of their measurements was more precise?**

A 4.5 L **C** 4.25 L
B 4 L **D** 4,259 mL

Name ______________________

Practice Objective 33

DIRECTIONS

Read each question and choose the most appropriate measure. Then mark the space for the answer you have chosen.

SAMPLE

The width of a school locker

A 14 in. C 11 ft

B 42 in. D $2\frac{1}{2}$ yd

1. **The height of a toaster**

A 7 ft C 1 yd

B 7 in. D 2 ft

2. **The height of a doorknob**

F 20 in. H 5 ft

G 3 ft J 2 yd

3. **The height of a woman**

A 12 ft C 100 in.

B 1 yd D 5 ft

4. **The distance from Los Angeles to Atlanta**

F 2,200 in. H 2,200 mi

G 5,000 ft J 5,000 mi

5. **The height of a desk**

A 14 in. C 20 ft

B $2\frac{1}{2}$ ft D 14 yd

6. **The length of a tube of toothpaste**

F 75 in. H 6 in.

G 2 ft J 2 yd

7. **The length of a river**

A 300 mi C 400 yd

B 1,000 in. D 150 ft

8. **The length of a toothpick**

F 2 ft H 20 in.

G 40 yd J 2 in.

9. **The length of a driveway from the curb to the garage**

A 17 in. C 1,700 ft

B 17 yd D 170 mi

10. **The length of a basketball court**

F 28 in. H 28 yd

G 28 ft J 2.8 mi

11. **Each of four students measured the length of their classroom. Which of their measurements was more precise?**

A 4 yd C 3 yd

B 3.5 yd D $11\frac{1}{2}$ ft

Name ______________________________

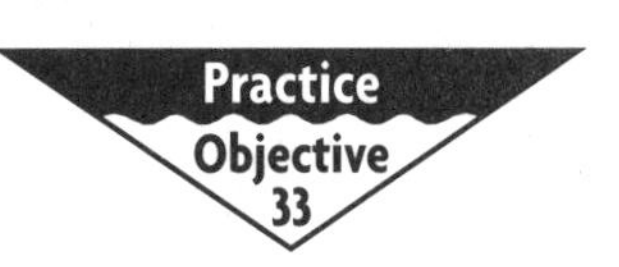

DIRECTIONS

Read each question and choose the most appropriate measure.
Then mark the space for the answer you have chosen.

SAMPLE

The weight of a turkey

A 12 lb C 45 lb
B 20 oz D 60 oz

1 The weight of a sack of potatoes

A 2 oz C 5 lb
B 25 oz D 500 lb

2 The weight of a bicycle

F 200 lb H 2 lb
G 20 lb J 20 oz

3 The weight of a pumpkin

A 1 ton C 200 lb
B 12 lb D 20 oz

4 The weight of a truck

F 10 T H 1,000 oz
G 100 lb J 10 lb

5 The capacity of a coffee mug

A 9 fl oz C 2 pt
B 30 fl oz D 10 cups

6 The capacity of a vegetable serving bowl

F 1 gal H 3 pt
G 10 qt J $\frac{1}{2}$ cup

7 The capacity of a juice box

A 8 pt C 8 fl oz
B 1 qt D 1 fl oz

8 The capacity of a milk pitcher

F 10 qt H 20 fl oz
G 20 qt J 2 fl oz

9 The capacity of a kitchen sink

A 50 qt C 10 cups
B 4 pt D 1 gal

10 The capacity of a can of soda

F 2 cups H 1.75 gal
G 2.5 qt J 3 gal

11 Enid and three friends measured the weight of a cocker spaniel. Which of their measurements was more precise?

A 672 oz C 42 lb
B 672.5 oz D 42.2 lb

Name ______________________________

Practice Objective 34

DIRECTIONS

Read each question and choose the best answer. Then mark the space for the answer you have chosen. If a correct answer is *not here,* mark the space for NH.

SAMPLE

150 cm =

A 15 m
B 0.15 mm
C 1.5 m
D 1.5 km
E NH

1 **62 cm =**

A 620 mm
B 6,200 m
C 6.2 km
D 0.062 m
E NH

2 **158 cm =**

F 1.58 mm
G 1.58 m
H 0.158 m
J 1.58 km
K NH

3 **34 mm =**

A 3.4 cm
B 3.4 m
C 0.34 m
D 0.34 km
E NH

4 **2,340 m =**

F 0.234 km
G 2.34 km
H 23.4 km
J 234 km
K NH

5 **8,070 mL =**

A 8.07 mL
B 807 L
C 80,700 mL
D 80.7 L
E NH

6 **6.1 L =**

F 61 mL
G 610 mL
H 6,100 mL
J 61,000 mL
K NH

7 **7.6 kg =**

A 7,600 g
B 760 g
C 7,600 mg
D 760 mg
E NH

8 **2.8 g =**

F 2,800 kg
G 28 kg
H 280 mg
J 2,800 mg
K NH

9 **A can of beans weighs 425 grams. How many kilograms will 24 cans weigh?**

A 4.25 kg
B 42.5 kg
C 10.2 kg
D 10,200 kg
E NH

Name ________________________________

Practice Objective 34

DIRECTIONS

Read each question and choose the best answer. Then mark the space for the answer you have chosen. If a correct answer is *not here,* mark the space for NH.

SAMPLE

144 in. =

A 12 yd
B 6 yd
C 12 ft
D 6 ft
E NH

1 **10 mi** =

A 5,280 ft
B 10,000 ft
C 52,800 ft
D 528,000 ft
E NH

2 **1,760 yd** =

F 5,280 ft
G 5,280 in.
H 176 ft
J $146\frac{2}{3}$ ft
K NH

3 **252 in.** =

A 9,072 yd
B 3,024 ft
C 21 ft
D 7 ft
E NH

4 **51 ft** =

F 612 yd
G 153 yd
H 612 in.
J 17 in.
K NH

5 **10 qt** =

A 20 gal
B 5 gal
C 20 c
D 5 pt
E NH

6 **28 qt** =

F 7 gal
G 4 gal
H 14 pt
J 56 c
K NH

7 **16 gal** =

A 32 qt
B 32 pt
C 64 qt
D 264 pt
E NH

8 **8 lb** =

F 8 oz
G 64 oz
H 96 oz
J 128 oz
K NH

9 **Tasha wants to make 160 cups of punch for her party. How many quarts of punch will she make?**

A 40 qt
B 80 qt
C 320 qt
D 640 qt
E NH

Name ______________________________

Practice Objective 35

DIRECTIONS

Read each question and choose the best approximate answer. Then mark the space for the answer you have chosen. If a correct answer is *not here,* mark the space for NH.

SAMPLE

80 in. ≈ ■ m

A 20
B 7
C 2
D 1
E NH

1 **5 m ≈ ■ yd**

A 2
B 5
C 15
D 500
E NH

2 **9 ft ≈ ■ m**

F 90
G 27
H 3
J 2
K NH

3 **2 mi ≈ ■ km**

A 1
B 4
C 8
D 20
E NH

4 **$\frac{1}{2}$ in. ≈ ■ cm**

F 12
G 4
H 2
J 1
K NH

5 **8 yd ≈ ■ m**

A $\frac{2}{3}$
B 16
C 24
D 96
E NH

6 **6 qt ≈ ■ L**

F 12
G 6
H 3
J 1.5
K NH

7 **1 gal ≈ ■ L**

A 16
B 12
C 8
D 4
E NH

8 **3 c ≈ ■ mL**

F 75
G 750
H 900
J 1,500
K NH

9 **A specialty can of coffee weighs 2 pounds. About how many kilograms does it weigh?**

A About 1 kg
B About 2.5 kg
C About 3 kg
D About 4 kg

Name ______________________________

DIRECTIONS

Read each question and choose the best answer. Then mark the space for the answer you have chosen.

SAMPLE

Estimate the area. Each square is 1 square centimeter (cm^2).

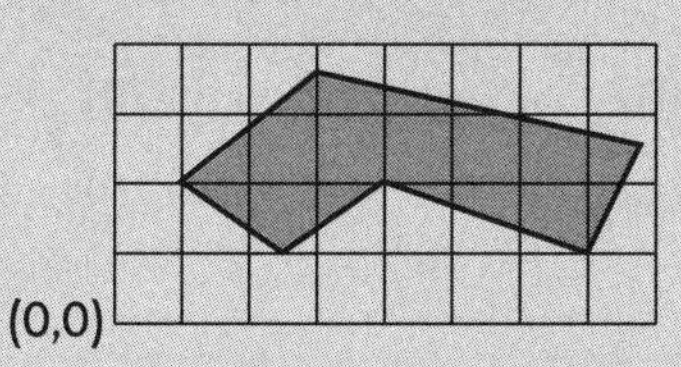

A About 4 cm^2
B About 8 cm^2
C About 12 cm^2
D About 21 cm^2

Use the figure for questions 1 and 2. Each square is 1 square centimeter.

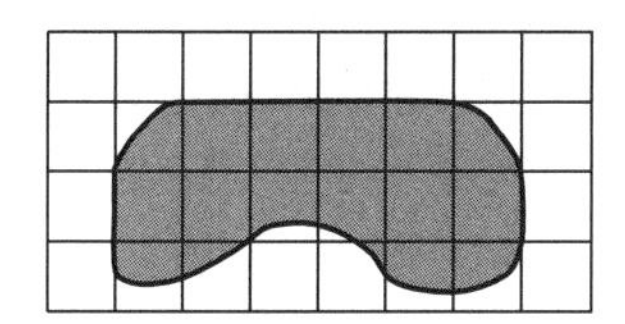

1 Estimate the perimeter.

A About 20 cm
B About 18 cm
C About 14 cm
D About 10 cm

2 Estimate the area.

F About 13 cm^2
G About 16 cm^2
H About 20 cm^2
J About 25 cm^2

Use the figure for questions 3 and 4. Each square is 1 cm^2.

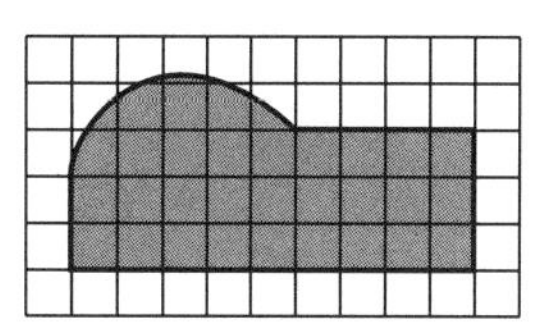

3 Estimate the perimeter.

A About 15 cm
B About 20 cm
C About 24 cm
D About 30 cm

4 Estimate the area.

F About 26 cm^2
G About 30 cm^2
H About 33 cm^2
J About 45 cm^2

Alfonso drew this design for art class. Use the design for questions 5 and 6. Each square is 1 cm^2.

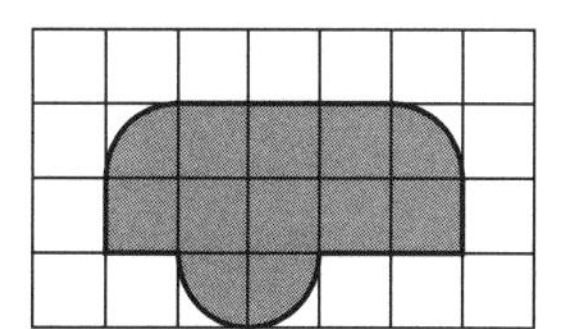

5 Estimate the perimeter.

A About 8 cm
B About 12 cm
C About 16 cm
D About 25 cm

6 Estimate the area.

F About 15 cm^2
G About 14 cm^2
H About 11 cm^2
J About 9 cm^2

Name ______________________________

DIRECTIONS

Read each question and choose the best answer. Then mark the space for the answer you have chosen.

Use your centimeter ruler and this map to help you answer the sample question and question 1.

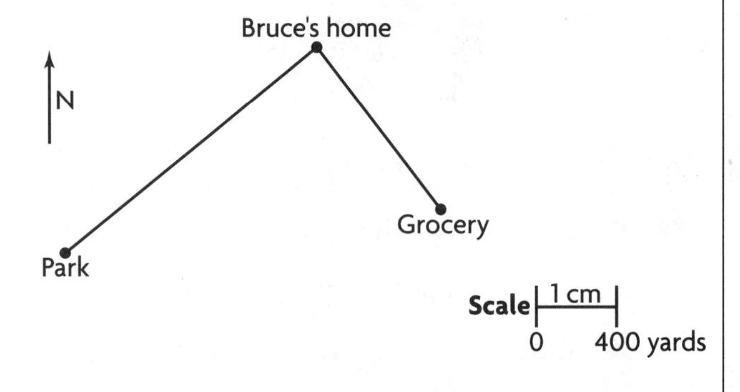

SAMPLE

What is the actual distance from Bruce's home to the park?

A 1,600 yd C 800 yd

B 1,200 yd D 400 yd

1 What is the actual distance from Bruce's home to the grocery?

A 400 yd C 800 yd

B 600 yd D 1,000 yd

2 The distance from Tuan's city to a neighboring city is 4 inches on a map. What is the actual distance if the scale is 1 in.:20 mi?

F 80 mi H 50 mi

G 60 mi J 5 mi

3 The distance from Mel's home to Hanover measures $5\frac{1}{2}$ inches on a scale drawing. What is the actual distance if the scale is 1 in.:100mi?

A 505 mi C 555 mi

B 550 mi D 650 mi

4 A tiny screw used in eyeglass frames measures 4 millimeters in length. What is the length of the screw in a scale drawing if the scale is 2 cm:1 mm?

F 4 cm H 8 cm

G 6 cm J 10 cm

5 The actual length of a bird feeder is 10 inches. What is the length of the bird feeder in a scale drawing if the scale is 1 in.:4 in.?

A $1\frac{1}{2}$ in. C 5 in.

B $2\frac{1}{2}$ in. D 10 in.

6 A school bus made a round trip from Chester to Richmond for a soccer game. The map distance between Chester and Richmond is 4 in. What is the actual round-trip distance if the scale is 1 in. = 3 mi?

F 12 mi H 24 mi

G 18 mi J 36 mi

Name ___________________________________

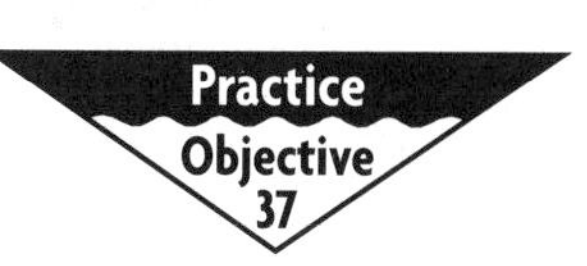

DIRECTIONS

Read each question and choose the best answer. Then mark the space for the answer you have chosen.

SAMPLE

The length-to-width ratio of a scale drawing is ___?___ the length-to-width ratio of the original object.

A Larger than
B Smaller than
C The same as
D Twice

1 **Which of the following would be the best scale to use if you are making a scale drawing of a new car?**

A 5 m:1 cm
B 1 in.:2 ft
C 2 ft:1 in.
D 1 in.:12 ft

2 **Jamie wants to make a scale drawing of a beetle with an actual length of 3 centimeters. She decides to use 12 centimeters for the length in the scale drawing. What scale should she use?**

F 1 in.:12 in.
G 1 in.:4 in.
H 1 cm:4 cm
J 4 cm:1 cm

3 **Arafat uses 10 centimeters to represent the width of a flower in a scale drawing. The actual width is 2 centimeters. What scale should he use?**

A 1 cm:10 cm
B 5 cm:1 cm
C 2 cm:10 cm
D 5 cm:2 cm

4 **Maribelle decides to use 1 cm:4 m for a scale drawing of a house. The actual length is 20 meters. How long should Maribelle make the length of the house in the scale drawing?**

F 5 centimeters
G 8 centimeters
H 10 centimeters
J 15 centimeters

5 **Ken decides to use a scale of 5 in.:2 in. for a scale drawing of a beetle. The actual length of the beetle is 3 inches. How long should Ken make the length in the drawing?**

A 6 inches
B 7 inches
C $7\frac{1}{2}$ inches
D 15 inches

6 **Yasmeen makes a map using a scale of 1 in.:50 mi. If one of the map distances is 4 inches, what does n represent in the proportion $\frac{1}{50} = \frac{4}{n}$?**

F The actual distance in miles
G The map distance in inches
H The straight-line distance in inches
J The scale

Name ______________________________

Practice Objective 38

DIRECTIONS

Read each question and choose the best answer. Then mark the space for the answer you have chosen.

SAMPLE

Use your centimeter ruler to help you answer the question. What is the perimeter of this triangular sticker?

A 10.5 cm C 7 cm

B 7.5 cm D 6.5 cm

1 Use your centimeter ruler to help you answer the question. What is the perimeter of this rectangular name tag?

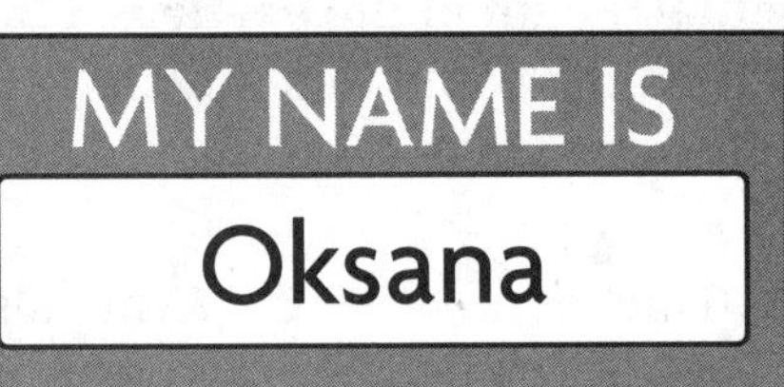

A 8.5 cm C 16 cm

B 11 cm D 17 cm

2 The city's rectangular parking lot is 120 yd long and 90 yd wide. How much fencing is needed to enclose the parking lot?

F 210 yd H 480 yd

G 420 yd J 840 yd

3 Max put up a fence around this garden to keep small animals out. What is the perimeter of this vegetable garden?

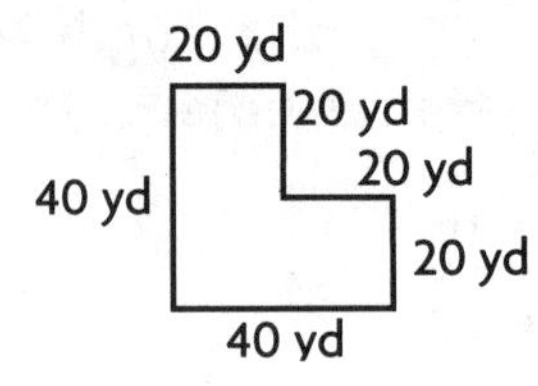

A 120 yd C 160 yd

B 140 yd D 180 yd

4 The perimeter of this town is 103 miles. What is the missing length?

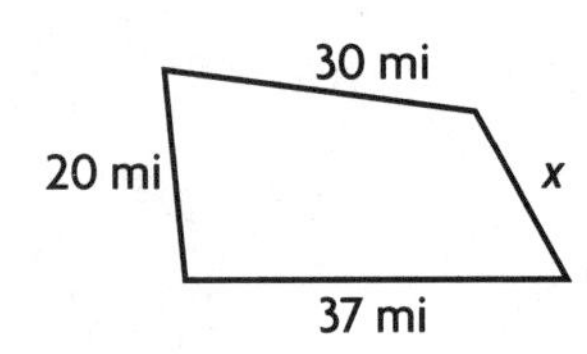

F 26 mi H 20 mi

G 22 mi J 16 mi

5 A puzzle is in the shape of a regular hexagon. The perimeter of the puzzle is 48 in. What is the length of each side?

A 6 in. C 12 in.

B 8 in. D 16 in.

6 The perimeter of a rectangular door is 20 ft. The door is 3 ft wide. How tall is the door?

F 10 ft H 7 ft

G 8.5 ft J 3 ft

Name ______________________________

DIRECTIONS

Read each question and choose the best answer. Then mark the space for the answer you have chosen.

SAMPLE

Use your centimeter ruler to help answer this question. Which stamp has an area of 4 cm^2?

A 

C 

B 

D 

1 **Use your centimeter ruler to help answer this question. Which eraser has an area of 3 cm^2?**

A 

C 

B 

D 

2 **Marcus drew a square that measured 5 inches on each side. If he doubles the length of each side, what will the area of the square be?**

F 10 in.^{2} H 25 in.^{2}

G 20 in.^{2} J 100 in.^{2}

3 **Leroy has a rectangular garden that is 20 feet by 12 feet. He decides to double the garden's area. Which of the following dimensions will double the area of his garden?**

A 30 ft by 12 ft C 40 ft by 24 ft

B 40 ft by 12 ft D 80 ft by 24 ft

4 **Lena built a square table that measured 4 feet on each side. She used 16 square tiles to cover the top of the table. She then built another square table that measured 2 feet on each side. If she uses the same kind of tiles to cover the top of the table, how many tiles will she need?**

F 14 tiles H 8 tiles

G 12 tiles J 4 tiles

5 **Corey is constructing a 12-foot by 3-foot rectangular walkway made of red bricks. If each brick has an area of 1 square foot and costs $3.50, what is the total cost of the bricks?**

A $105.00 C $126.00

B $108.00 D $133.00

6 **Kendra bought 36 carpet tiles to carpet her rectangular office. Michaelene's office is twice as long and twice as wide as Kendra's office. How many carpet tiles will be needed to carpet Michaelene's office?**

F 18 tiles H 90 tiles

G 72 tiles J 144 tiles

Name ______________________________

DIRECTIONS

Read each question and choose the best answer. Then mark the space for the answer you have chosen.

SAMPLE

Find the perimeter of the square. Use the formula $P = 4s$.

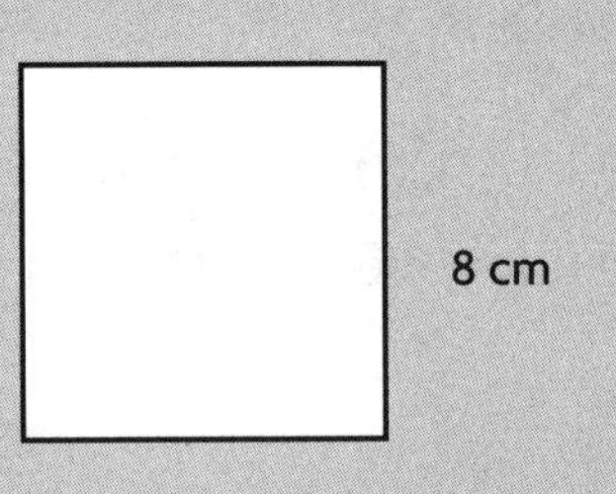

A 8 cm
B 16 cm
C 32 cm
D 64 cm

1 **Find the perimeter of the square. Use the formula $P = 4s$.**

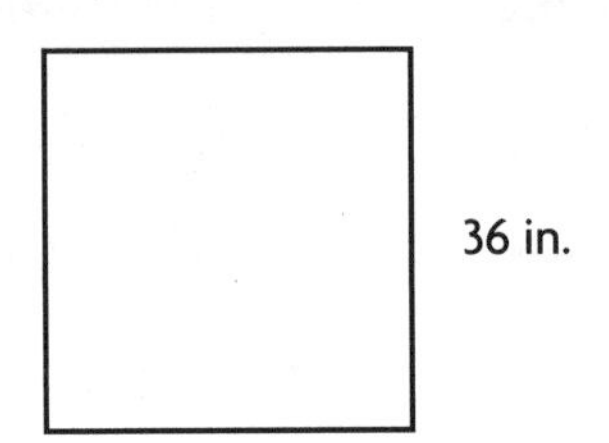

A 9 in.
B 72 in.
C 144 in.
D 1,296 in.

2 **Find the perimeter of the rectangle. Use the formula $P = (2 \times l) + (2 \times w)$.**

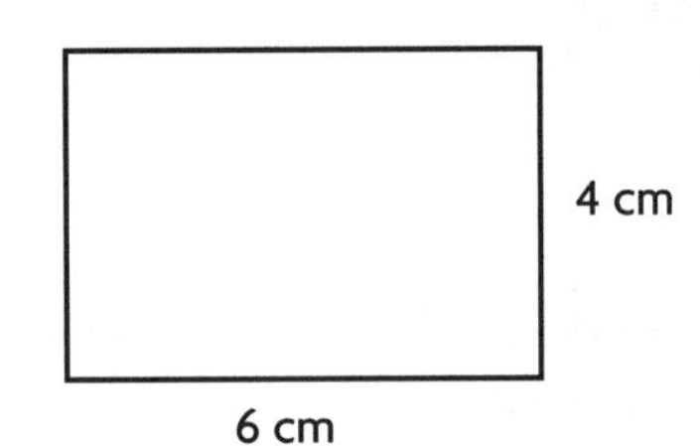

F 10 cm
G 20 cm
H 24 cm
J 48 cm

3 **Find the perimeter of the regular hexagon. Use the formula $P = 6s$.**

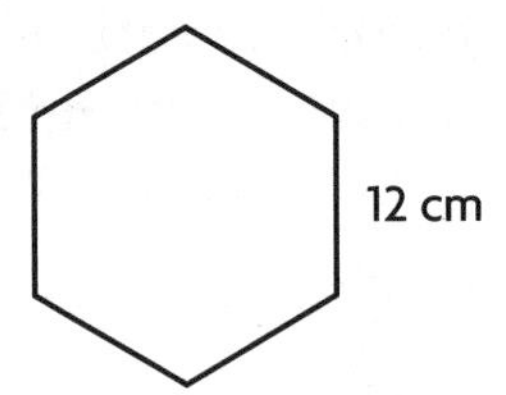

A 48 cm
B 72 cm
C 82 cm
D 144 cm

4 **Find the circumference of the circle. Use the formula $C = \pi d$. Use 3.14 for π.**

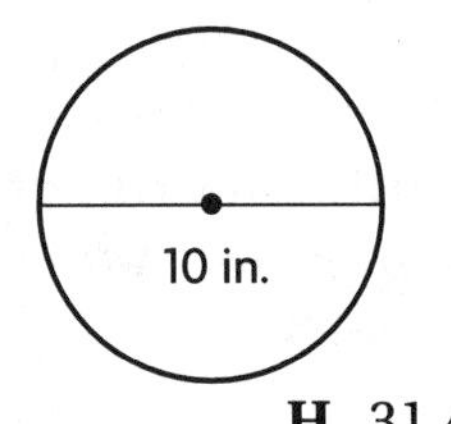

F 3.14 in.
G 30.14 in.
H 31.4 in.
J 314 in.

5 **Find the circumference of the circle. Use the formula $C = \pi d$. Use $\frac{22}{7}$ for π.**

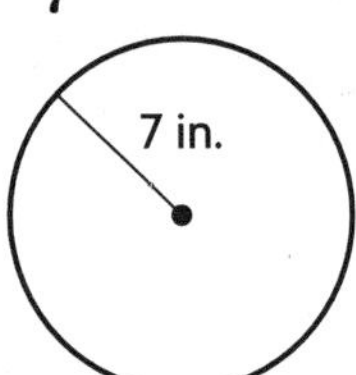

A 14 in.
B 22 in.
C 44 in.
D 88 in.

Name ______________________

Practice Objective 39

DIRECTIONS

Read each question and choose the best answer. Then mark the space for the answer you have chosen.

SAMPLE

Find the area of the square. Use the formula $A = s^2$

5 cm

A 10 cm^2 C 25 cm^2

B 20 cm^2 D 125 cm^2

1 **Find the area of the rectangle. Use the formula $A = lw$.**

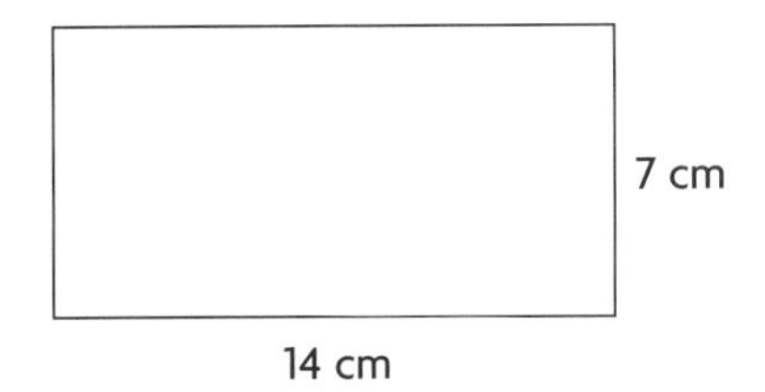

A 196 cm^2 C 42 cm^2

B 98 cm^2 D 21 cm^2

2 **Find the area of the parallelogram. Use the formula $A = bh$.**

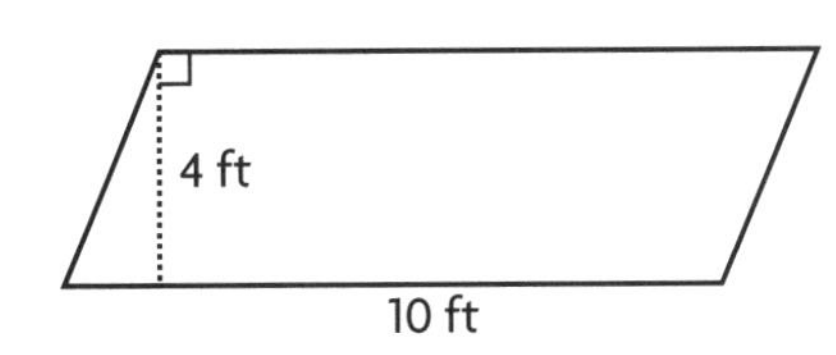

F 20 ft^2 H 40 ft^2

G 28 ft^2 J 80 ft^2

3 **Find the area of the circle. Use the formula $A = \pi r^2$. Use 3.14 for π. Round the answer to the nearest whole unit.**

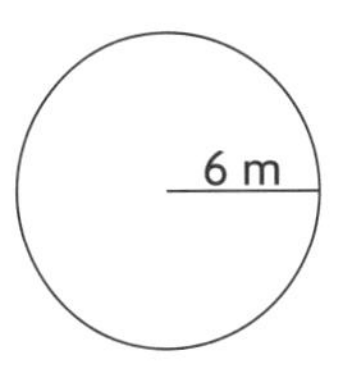

A About 36 m^2 C About 113 m^2

B About 108 m^2 D About 144 m^2

4 **Find the volume of the prism. Use the formula $V = lwh$.**

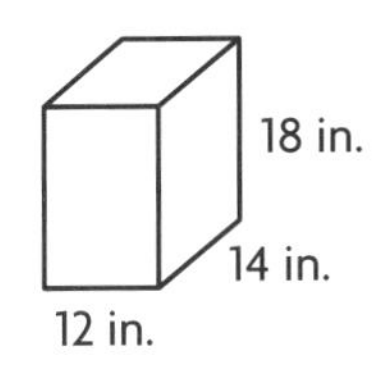

F 44 in.^3 H $1{,}512 \text{ in.}^3$

G 186 in.^3 J $3{,}024 \text{ in.}^3$

5 **Find the volume of the cylinder. Use the formula $V = \pi r^2 h$. Use 3.14 for π. Round the answer to the nearest whole unit.**

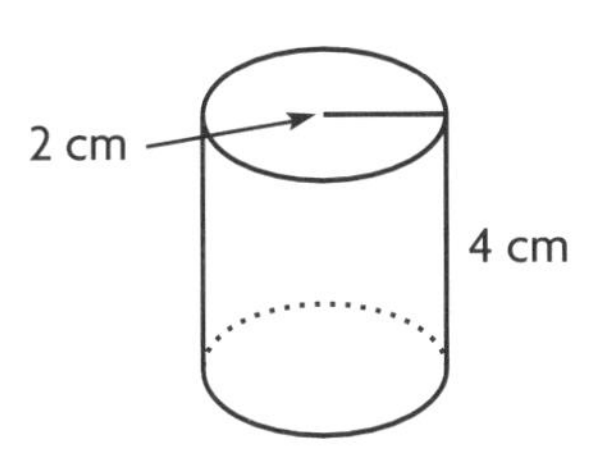

A 16 cm^3 C 32 cm^3

B 25 cm^3 D 50 cm^3

Name ______________________________

DIRECTIONS

Read each question and choose the best answer. Then mark the space for the answer you have chosen.

SAMPLE

Based on the following statement, which statement is true?

If a number has two digits, then it is less than 100.

A All numbers less than 100 have two digits.

B No numbers less than 100 have two digits.

C All two-digit numbers are less than 100.

D No two-digit numbers are less than 100.

1 **Based on the following statement, which statement is true?**

If an angle measures 50°, then it is an acute angle.

A All acute angles measure 50°.

B No acute angles measure 50°.

C All angles that measure 50° are acute.

D No 50° angles are acute.

2 **Based on the following statement, which statement is true?**

If a figure is a square, then it is a quadrilateral.

F All squares are quadrilaterals.

G No squares are quadrilaterals.

H All quadrilaterals are squares.

J No quadrilaterals are squares.

3 **Based on the following statement, which statement is true?**

If an integer is less than 0, then it is negative.

A No integers are negative.

B No negative numbers are integers.

C All negative numbers are integers.

D All integers less than 0 are negative.

4 **Based on the following statement, which statement is true?**

If a figure is a rectangle, then it is a parallelogram.

F All parallelograms are rectangles.

G No parallelograms are rectangles.

H All rectangles are parallelograms.

J No rectangles are parallelograms.

5 **Based on the following statement, which statement is true?**

If two lines are perpendicular, then they intersect.

A All intersecting lines are perpendicular.

B Two perpendicular lines intersect.

C No intersecting lines are perpendicular.

D No perpendicular lines intersect.

6 **Based on the following statement, which statement is true?**

If a figure is a square, then it it a rectangle.

F All rectangles are squares.

G All squares are rectangles.

H No squares are rectangles.

J No rectangles are squares.

Name ______________________________

DIRECTIONS

Read each question and choose the best answer. Then mark the space for the answer you have chosen.

For the sample question and question 1, use the stem-and-leaf plot of test scores for a sixth-grade class.

Class Test Scores

Stem	Leaves
5	8 9
6	0 4 7 8 8
7	0 1 3 4 6 7 9 9
8	0 0 1 2 2 4 5 6 6 8 9
9	1 3 5 6 7

SAMPLE

What score is shown by the fourth stem and its second leaf?

A 68 C 80

B 79 D 81

1 **Where should a score of 90 be placed on the stem-and-leaf plot?**

A The fourth stem and its eleventh leaf

B The fifth stem and its first leaf

C The fifth stem and its third leaf

D The fifth stem and its fifth leaf

2 **The histogram below shows the ages of people at a movie. What label is missing from the histogram?**

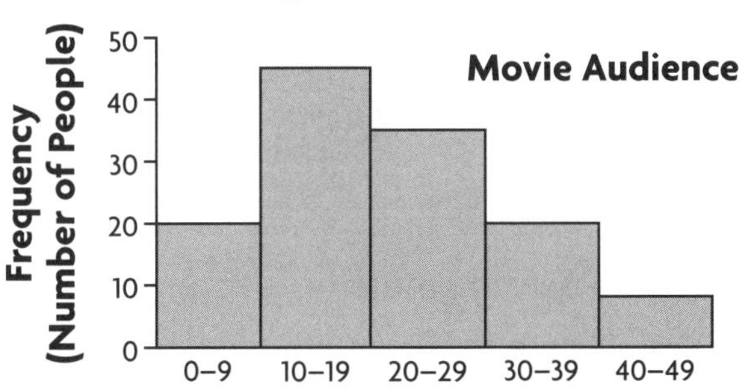

F Frequency H Audience

G Number of people J Age

3 **What would the histogram for the frequency table below look like?**

15-Mile Race	
Minutes	Runners
0 - 59	20
60 - 119	35
120 - 179	10
180 - 239	5

A

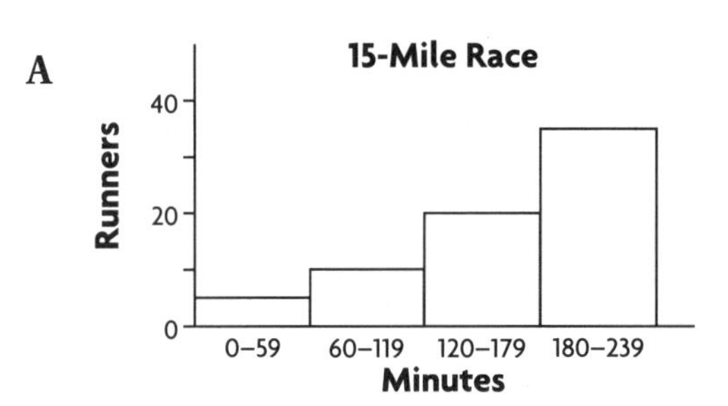

B

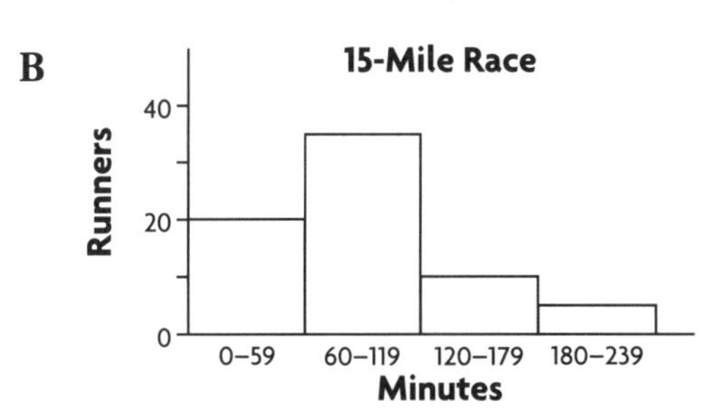

C

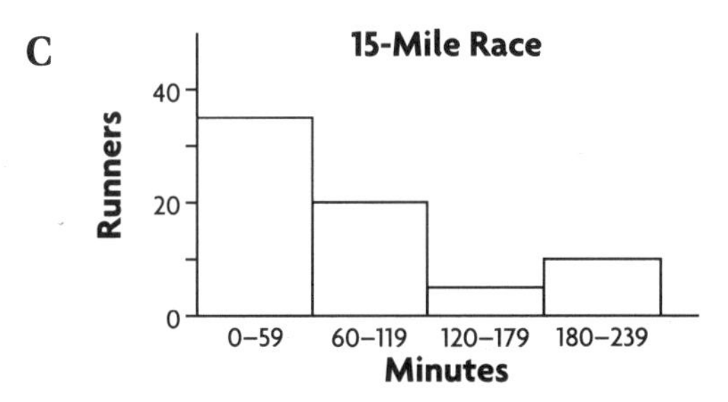

D

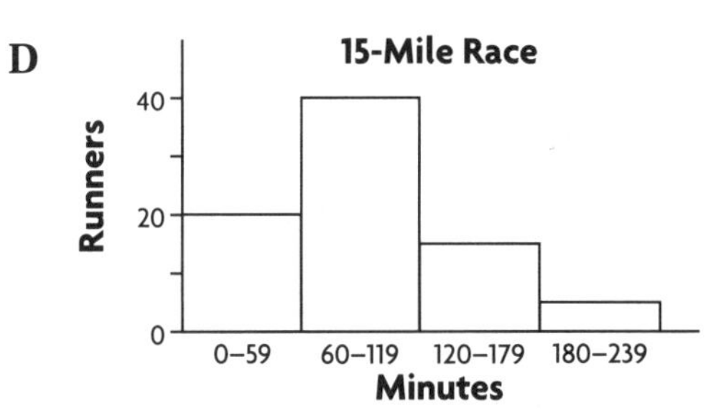

Name ____________________

DIRECTIONS

Read each question and choose the best answer. Then mark the space for the answer you have chosen.

For the sample question and question 1, use the table showing the results when students were surveyed about their favorite pizzas.

Favorite Pizza	
Plain	15
Mushrooms	35
Pepperoni	50

SAMPLE

How many students were surveyed?

A 50 students **C** 90 students

B 85 students **D** 100 students

1 **Barry is using the data about favorite pizzas to make a circle graph. What will Barry's graph look like?**

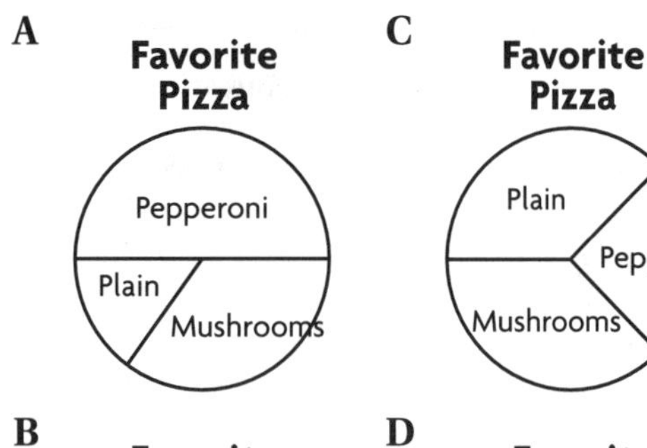

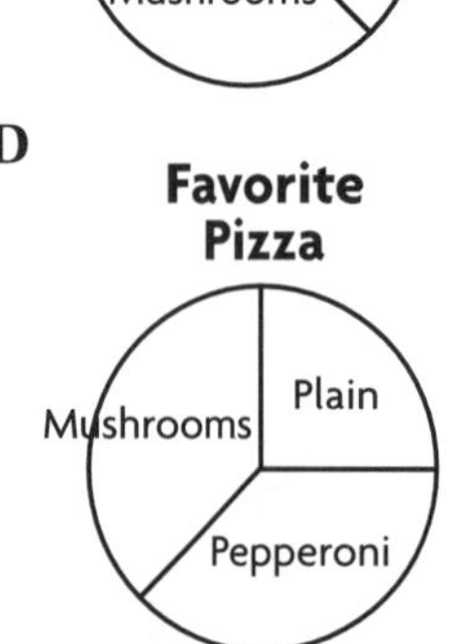

2 **The Schools in Northwest County conducted a survey of 100 students to find out if they liked the idea of changing the school schedule. What did the schools learn from the survey?**

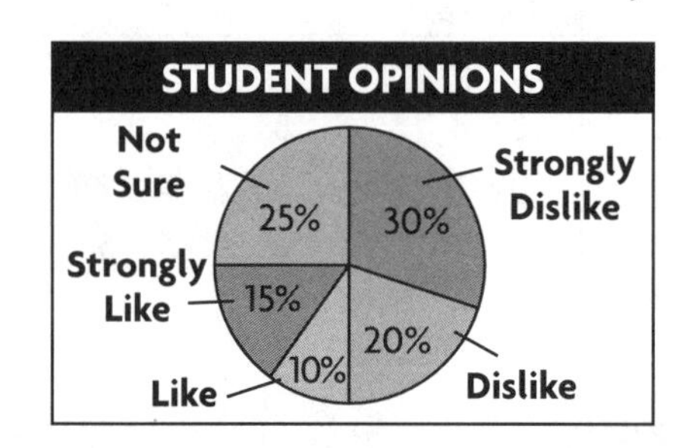

F Of the students surveyed, 25% favored the idea.

G Of the students surveyed, 60% disliked the idea.

H Opinions were evenly divided.

J More than half of the students surveyed were against the idea.

3 **Which statement about the 8th-grade band members is true?**

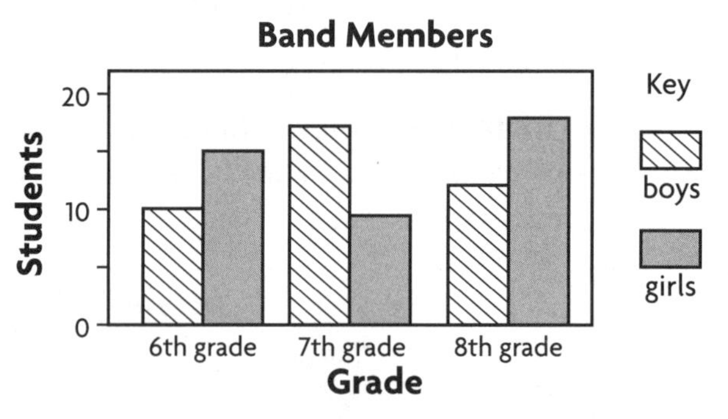

A There are more girls than boys.

B There are more boys than girls.

C There are fewer 8th graders than 7th graders.

D There are fewer 8th graders than 6th graders.

Name ______________________________

Practice Objective 41

DIRECTIONS

Read each question and choose the best answer. Then mark the space for the answer you have chosen.

For the sample question and questions 1 and 2, use the graph below.

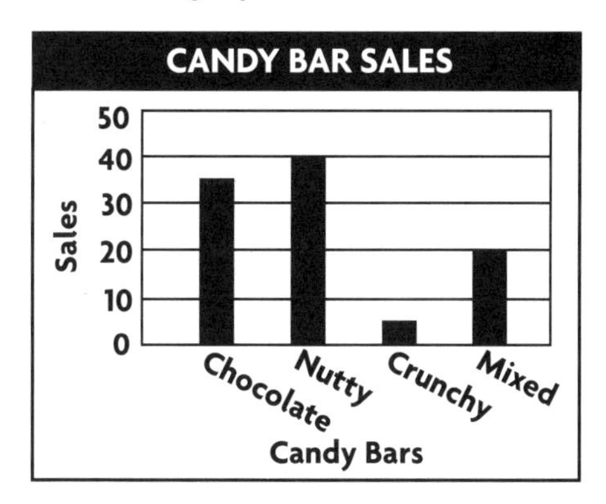

SAMPLE

Which kind of candy bar was the most popular?

A Chocolate C Crunchy
B Nutty D Mixed

1 How does the number of mixed candy bars sold compare with the number of crunchy candy bars sold?

A The same number of each was sold.
B Twice as many crunchy candy bars were sold.
C Twice as many mixed candy bars were sold.
D About four times as many mixed candy bars were sold.

2 The students plan to sell a fruit and nut candy bar at their next sale. They still want only four kinds of candy bars. Which candy bar should they stop selling?

F Chocolate H Crunchy
G Nutty J Mixed

3 What was the trend for sales of cars?

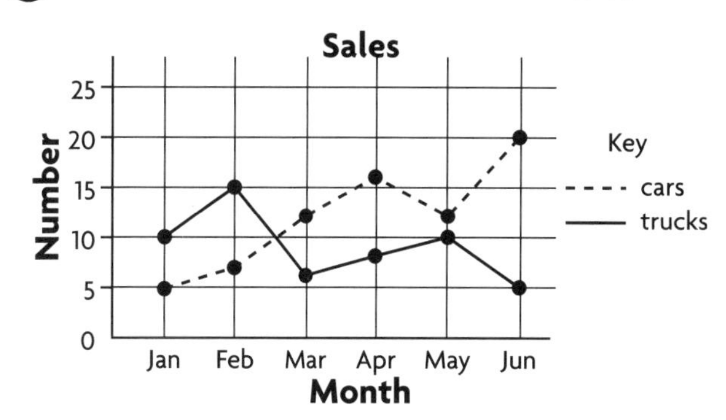

A Increased
B Decreased
C Stayed the same
D Less than truck sales

4 The enrollment at Shields Middle School was 105 in 1994, 110 in 1995, 115 in 1996, 120 in 1997, and 125 in 1998. If this data is graphed, what trend for enrollment would it show?

F Decreasing H Staying the same
G Increasing J Not enough data

5 According to the graph below, about how many times greater are the sales in Week 4 than the sales in Week 1?

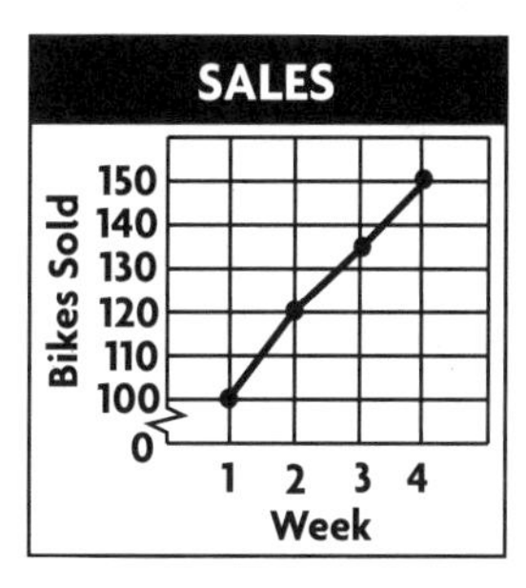

A About $1\frac{1}{2}$ times greater
B About twice as great
C About three times as great
D About five times as great

Name ______________________________

Practice Objective 42

DIRECTIONS

Read each question and choose the best answer. Then mark the space for the answer you have chosen.

SAMPLE

What is the mean (average) for this set of data?

20, 24, 27, 1, 23

A 18 **C** 20

B 19 **D** 26

1 **What is the mean (average) for this set of data?**

51, 48, 39, 53, 47, 44, 42, 55, 44

A 46 **C** 47

B 46.5 **D** 47.5

2 **What is the mode for this set of data?**

14, 21, 17, 25, 20, 17

F 17 **H** 18.5

G 18 **J** 19

3 **What is the mode for this set of data?**

11, 16, 14, 18, 18, 13

A 11 **C** 15

B 16 **D** 18

4 **What is the median for this set of data?**

49, 52, 58, 49, 56

F 58 **H** 52

G 52.8 **J** 49

5 **What is the median for this set of data?**

45, 58, 37, 29, 76, 19, 68, 91, 57

A 45 **C** 54.2

B 47 **D** 57

6 **What is the range for this set of data?**

34, 47, 38, 42, 15, 59, 71

F 15 **H** 56

G 50 **J** 71

7 **What is the range for this set of data?**

11, 15, 13, 12, 15, 13, 12, 14, 13, 15, 15, 11, 16, 12, 11, 12

A 5 **C** 11

B 6 **D** 12

Name ______________________

Practice Objective 42

DIRECTIONS

Read each question and choose the best answer. Then mark the space for the answer you have chosen.

SAMPLE

A figure skater was awarded these scores by six judges: 7.5, 8.5, 8.0, 8.2, 8.3, and 7.5. What is the mean (average) score?

A 8.5 **C** 8.0
B 8.1 **D** 7.5

1 **Dan scored these points during the basketball season: 14, 21, 17, 25, 17, 18, 16, 22, 17, and 23. What is the mean number of points Dan scored?**

A 11 **C** 19
B 17 **D** 25

2 **Joanne bowled three games and averaged 134 points per game. If she bowled 125 for the first game and 145 for the second game, what was her score for the third game?**

F 402 **H** 132
G 279 **J** 123

3 **During May, the median house sale in the Carol Stream district was $134,095. There were 245 sales made in May. How many sales were below the median sale price?**

A 122 sales **C** 245 sales
B 123 sales **D** 547 sales

For questions 4 and 5, use the line plot. It shows scores on a math test taken by 19 students.

Test Scores on a Math Test

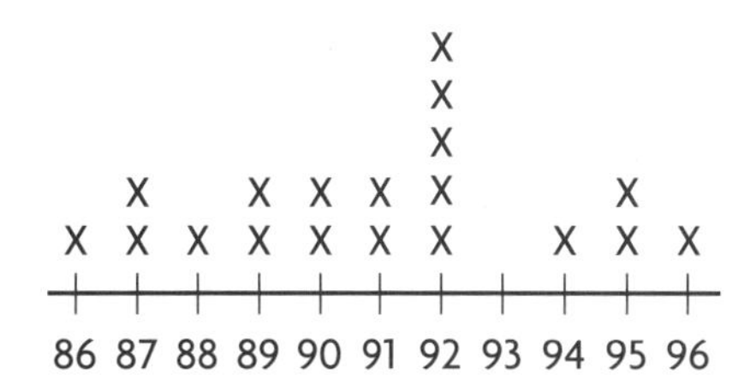

4 **What is the mode score?**

F 13 **H** 91
G 90.9 **J** 92

5 **What is the median score?**

A 13 **C** 91
B 90.9 **D** 92

6 **What is the range for the ages of musicians in a community orchestra?**

Ages of Musicians

20	52	42	59	46
48	64	36	21	40
14	45	38	29	34
39	35	61	19	27
26	18	52	57	

F 64 **H** 38.5
G 50 **J** 14

Name ______________________

DIRECTIONS

Read each question and choose the best answer. Then mark the space for the answer you have chosen.

SAMPLE

Kelly spins the spinner two times. What are the possible outcomes?

A 1-1, 1-2, 1-3, 2-1, 2-2, 2-3, 3-1, 3-2, 3-3

B 1-1, 1-2, 1-3, 2-2, 2-3, 3-3

C 2-2, 2-3, 3-3, 3-2, 3-1

D 1-1, 1-2, 1-3, 2-1, 2-2, 2-3, 4-1, 4-2, 4-3

1 **Denise spins this spinner once and tosses a coin once. What are the possible outcomes?**

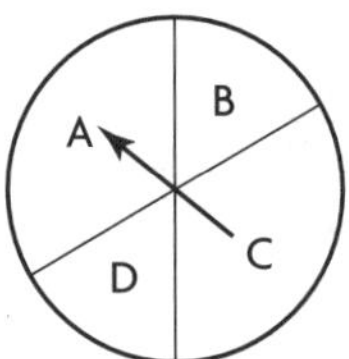

A 2 outcomes C 6 outcomes

B 4 outcomes D 8 outcomes

2 **Cari made a tree diagram to find the outcomes if she tosses a penny and a nickel. What possible outcome is missing?**

Penny	Nickel	Outcome
Heads	Heads	Heads, heads
	Tails	Heads, tails
Tails	Heads	Tails, heads
	Tails	

F Heads (penny), Tails (nickel)

G Heads (penny), Heads (nickel)

H Tails (penny), Tails (nickel)

J Tails (penny), Heads (nickel)

3 **Kendra tosses a number cube labeled from 1 to 6 and then tosses a coin. What are the possible outcomes?**

A 12 outcomes C 6 outcomes

B 10 outcomes D 2 outcomes

4 **Mike tosses two number cubes each labeled from 1 to 6. What are the possible outcomes?**

F 6 outcomes H 36 outcomes

G 12 outcomes J 72 outcomes

5 **If you spin the spinner twice, which outcome is not possible?**

A A sum of 10 C A sum of 18

B A sum of 14 D A sum of 4

6 **If you spin the spinner twice, which outcome is not possible?**

F A product of 3 H A product of 9

G A product of 6 J A product of 12

Name ______________________________

DIRECTIONS

Read each question and choose the best answer. Then mark the space for the answer you have chosen.

SAMPLE

Karl has 1 quarter, 6 dimes, 4 nickels, and 2 pennies in his pocket. If a coin falls out when he sits down, what is the probability that it will be a penny?

A $\frac{1}{2}$ C $\frac{2}{11}$

B $\frac{1}{3}$ D $\frac{2}{13}$

1 **Jamie has 1 red marble, 3 white marbles, 4 blue marbles, and 2 green marbles in a pouch. If one is removed without looking, what is the probability that it will be red?**

A 0.1 C 0.75

B 0.11 D 0.33

2 **Kala has 3 red, 1 black, 2 green, and 2 blue T-shirts. She chose a shirt without looking. What is the probability that she chose a red T-shirt?**

F $\frac{3}{4}$ H $\frac{1}{4}$

G $\frac{3}{8}$ J $\frac{1}{8}$

3 **There are 4 answer choices for a problem-solving test question. What is the probability you can guess the correct answer?**

A 0.1 C 0.2

B 0.17 D 0.25

4 **Suppose Jesse tosses a number cube labeled 1, 3, 5, 7, 9, 11. What is P(5)?**

F $\frac{1}{2}$ H $\frac{1}{4}$

G $\frac{1}{3}$ J $\frac{1}{6}$

5 **Suppose Sandy tosses a number cube labeled from 3 to 8. What are Sandy's chances of rolling a 1 or a 2?**

A 0 C $\frac{1}{2}$

B $\frac{1}{3}$ D 1

6 **Use the spinner. What is P(2 or 4)?**

F 0.25 H 0.60

G 0.50 J 0.75

7 **What is the probability of hitting the shaded section of the dart target?**

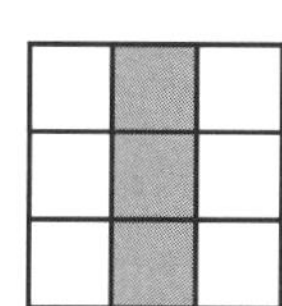

A $\frac{1}{4}$ C $\frac{1}{8}$

B $\frac{1}{3}$ D $\frac{1}{9}$

Name ______________________________

Practice Objective 45

DIRECTIONS

Read each question and choose the best answer. Then mark the space for the answer you have chosen.

SAMPLE

David has a pair of black pants and a pair of brown pants. He also has a white shirt, a tan shirt, and a red shirt. How many different outfits can he make using one pair of pants and one shirt?

A 2 outfits

B 3 outfits

C 5 outfits

D 6 outfits

1. **Maxine has a black skirt, a brown skirt, and a navy skirt. She also has a white blouse, a red blouse, and a yellow blouse. How many different outfits can she make using one skirt and one blouse?**

A 6 outfits

B 8 outfits

C 9 outfits

D 12 outfits

2. **Mona buys rye, white, wheat and pumpernickel bread. She also buys ham and roast beef. How many different kinds of sandwiches can she make using one type of bread and one type of meat?**

F 4 kinds

G 6 kinds

H 8 kinds

J 10 kinds

3. **Jim, Alex, Tom, Jack, and Bob play on a volleyball team. Two of these players were chosen to be co-captains. How many different pairs of co-captains are possible?**

A 5 pairs

B 8 pairs

C 10 pairs

D 15 pairs

4. **Carla, Dona, and Ruth volunteered to help Mr. Marshall. In how many ways can Mr. Marshall choose one person to record attendance and one person to take lunch money?**

F 2 ways

G 3 ways

H 5 ways

J 6 ways

5. **Pat has four new CDs. How many possible ways can he play the CDs if he always plays the same CD first?**

A 3 possible ways

B 6 possible ways

C 12 possible ways

D 18 possible ways

Name ___________________________________

DIRECTIONS

Read each question and choose the best answer. Then mark the space for the answer you have chosen.

SAMPLE

Stuart's birthday is the day before Genevieve's birthday. Bobby's birthday is 4 days after Stuart's. Bobby's birthday is Dec. 31. When is Genevieve's birthday?

A Dec. 27
B Dec. 28
C Jan. 3
D Jan. 4

1 **John's birthday is 3 days before Lizzie's. Lizzie's birthday is 5 days after Lorie's birthday. Lorie's birthday is Feb. 21. When is John's birthday?**

A Feb. 21
B Feb. 23
C Feb. 24
D Feb. 26

2 **Emma served pizza at her birthday party. Katy and Lisa each ate twice as much as Emma. Mary ate as much as Vicky and Lisa together. Patty ate as much as Mary. Vicky ate half as much as Lisa. Emma ate $\frac{1}{6}$ of a pizza. If they ate all the pizzas, what is the least number of pizzas Emma started with?**

F 1 pizza
G 2 pizzas
H 3 pizzas
J 4 pizzas

For questions 3–6, use the map. Each side of the square equals one city block.

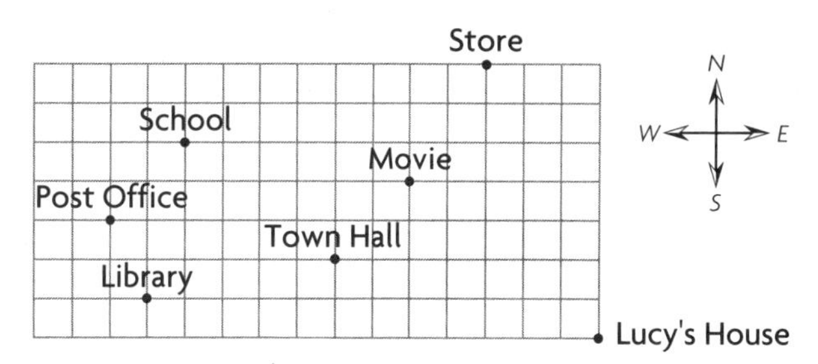

3 **Lucy left her house and walked 7 blocks north and 3 blocks west. Where was she?**

A Store
B School
C Movie
D Library

4 **Ariel left the library and went 3 blocks north and 7 blocks east. Where was she?**

F School
G Store
H Town Hall
J Movie

5 **Scott left school and walked 2 blocks south, 4 blocks east, and 1 block south. Where was he?**

A Movie
B Library
C Town Hall
D Store

6 **Lucy left home and biked 10 blocks west, 5 blocks north, and 1 block west. Where was she?**

F Store
G School
H Movie
J Library

Name ______________________________

Practice
Objective
46

DIRECTIONS

Read each question and choose the best answer. Then mark the space for the answer you have chosen.

SAMPLE

The sum of two numbers is 12. One number is $\frac{1}{2}$ of the other. What are the numbers?

A 4 and 8 **C** 3 and 6

B 7 and 5 **D** 9 and 3

1 **Which two numbers have a sum of 52, a difference of 10, and a product of 651?**

A 32 and 19 **C** 21 and 31

B 93 and 7 **D** 43 and 33

2 **The sum of two numbers is 0.62, and the difference is 0.24. What are the two numbers?**

F 0.36 and 0.12

G 0.40 and 0.22

H 0.43 and 0.19

J 0.44 and 0.18

3 **Describe the rule that best describes the following pattern: 0.7, 1.4, 2.8, 5.6, ...**

A Add 0.7.

B Add 1.4.

C Multiply by 2.

D Multiply by 0.2.

4 **Describe the rule that best describes the following pattern: $\frac{3}{8}, \frac{3}{4}, 1\frac{1}{8}, 1\frac{1}{2}, \ldots$**

F Divide by 2.

G Multiply by 2.

H Add $\frac{3}{16}$.

J Add $\frac{3}{8}$.

5 **To begin her exercise training, Kara walked for 20 minutes on the first day, 25 minutes on the second day, and 30 minutes on the third day. If she continues at the same rate, on which day will she walk for 60 minutes?**

A Day 6

B Day 7

C Day 8

D Day 9

6 **Andy began jogging for 12 minutes for a week. The next week he increased his jogging time to 16 minutes. The third week he increased his jogging time to 20 minutes. At this rate, during which week can he expect to jog more 30 minutes?**

F The 5th week

G The 6th week

H The 7th week

J The 10th week

Name ______________________________

Practice Objective 46

DIRECTIONS

Read each question and choose the best answer. Then mark the space for the answer you have chosen.

SAMPLE

The Randolph family bought $4\frac{1}{8}$ pounds of green apples and $2\frac{9}{10}$ pounds of red apples. About how many pounds of apples did the family buy altogether?

A About 6 lb
B About 7 lb
C About 9 lb
D About 10 lb

1 On a camping trip, the scout troop hiked $13\frac{5}{8}$ mi on the first day. They hiked $9\frac{1}{10}$ mi on the second day. About how much farther did they hike the first day than the second day?

A About $3\frac{1}{2}$ mi
B About $4\frac{1}{2}$ mi
C About $5\frac{1}{2}$ mi
D About 21 mi

2 Each bus seats no more than 65 passengers. Will 6 buses be able to seat 360 students?

F Yes, because $360 \div 6 < 65$.
G Yes, because $6 \times 65 = 360$.
H No, because $65 < 360 \div 6$.
J No, because $6 \times 65 > 360$.

3 Allen takes trumpet lessons. Lessons were available on Monday, Tuesday, Thursday, or Friday, at 4:00 P.M. or at 5:00 P.M. How many choices did he have?

A 4 choices
B 6 choices
C 8 choices
D 10 choices

4 Notebooks are available in green, red, blue, black, or yellow. They come with or without an inside pocket for papers. How many choices are there?

F 7 choices
G 10 choices
H 12 choices
J 15 choices

5 How many different kinds of sandwiches can be made using wheat, white, or rye bread, either turkey or ham, and with or without cheese?

A 6 kinds of sandwiches
B 8 kinds of sandwiches
C 10 kinds of sandwiches
D 12 kinds of sandwiches

6 How many different ways can you have $0.75 using only quarters, dimes, and/or nickels?

F 18 ways
G 14 ways
H 12 ways
J 10 ways

Name ______________________________

Practice Objective 46

DIRECTIONS

Read each question and choose the best answer. Then mark the space for the answer you have chosen.

SAMPLE

Mrs. Land ordered 108 tickets for the concert, including 95 student tickets. How many adult tickets did she order?

A 213 tickets

B 203 tickets

C 23 tickets

D 13 tickets

1 **The Lawrence family is planning a trip. If they travel at a rate of 60 miles per hour, how long will it take them to drive 270 miles?**

A $3\frac{1}{2}$ hours

B 4 hours

C $4\frac{1}{2}$ hours

D $4\frac{3}{4}$ hours

2 **Cindy is planning a trip. She plans to drive 180 miles the first day and $180 + y$ miles the second day. If $y = 60$, how far does she plan to drive the second day?**

F 60 miles

G 120 miles

H 240 miles

J 360 miles

3 **Sara is older than Sam by 4 years. Sara is 12 years old. If s = Sam's age, which equation would you use to find Sam's age?**

A $s + 4 = 12$

B $s - 4 = 12$

C $s = 12$

D $4 \times s = 12$

4 **After the first hour of the bake sale, 37 cupcakes were sold and 28 cupcakes were left. If t = the total number of cupcakes made for the sale, which equation would you use to find the total number made?**

F $t = 37 - 28$

G $t + 28 = 37$

H $t - 37 = 28$

J $t = 28$

5 **What is 90°F converted to degrees Celsius? Use the formula $C = \frac{5}{9} \times (F - 32)$. Round to the nearest degree.**

A 12°C

B 50°C

C 32°C

D 64°C

Name ______________________

Practice
Objective
47

DIRECTIONS

Read each question and choose the best answer. Then mark the space for the answer you have chosen. If a correct answer is *not here,* mark the space for NH.

SAMPLE

Use mental math to find $80 + 159 + 20$.

A 200 C 260 E NH
B 259 D 300

1 **Use mental math to find $70 + 150 + 130$.**

A 250 C 350 E NH
B 305 D 450

2 **Use mental math to find $63 - 38$.**

F 25 H 81 K NH
G 35 J 91

3 **Use mental math to find 64×4.**

A 266 C 246 E NH
B 256 D 224

4 **Use mental math to find $8 \times 10 \times 70$.**

F 560 H 5,600 K NH
G 5,400 J 54,000

5 **Use paper and pencil to find the remainder. $165 \div 7$**

A 4 C 21 E NH
B 5 D 1,155

6 **Use paper and pencil to find the product. 450×28**

F 1,260 H 12,600 K NH
G 126 J 126,000

7 **Use a calculator to find $\frac{5}{16}$ written as a decimal.**

A 31.25 C 0.8 E NH
B 0.03125 D 0.3125

8 **Use a calculator to find $3{,}901 \div 83$.**

F 47 H 407 K NH
G 470 J 4

9 **Anfernee earns $104.96 each day, including Saturdays and Sundays. At this rate, how much will he earn in a year? Which method is the most appropriate to answer the question?**

A Mental math
B Paper and pencil
C Calculator
D Computer

Name ______________________

Practice Objective 48

DIRECTIONS

Read each question and choose the best answer. Then mark the space for the answer you have chosen.

SAMPLE

Marty paid $6.38 for bread and milk at the food store. How much change did he receive from $10.00?

A $16.38 C $4.38

B $4.62 D $3.62

1 While shopping for a birthday gift for his grandmother, Ken paid $12.50 for a plant and $1.75 for a card. How much change did he receive from a twenty-dollar bill?

A $5.75 C $7.75

B $6.75 D $14.25

2 Elizabeth bought a sweater for $18.95 and a skirt for $24.95. Sales tax was $3.51. What was the total amount that she paid?

F $57.41 H $43.90

G $47.41 J $40.39

3 Mr. Clakin's dry cleaning bill is $43.75. He gave the clerk three twenty-dollar bills. How much change should he receive?

A $103.75 C $16.25

B $60.00 D $6.25

4 The regular price of a pair of in-line skates is $95.00. They were placed on sale for 40% off. How much is the discount?

F $57.00 H $47.50

G $38.00 J $19.00

5 The regular price of a VCR is $212.60. Jana received a discount of 25%. How much did she pay for the VCR?

A $53.15 C $159.45

B $101.30 D $175.00

6 A bicycle costs $129. The sales tax is 6%. How much is the sales tax?

F $7.74 H $136.74

G $77.40 J $206.40

7 Art supplies are priced at $35. Sales tax is 7%. What is the total cost of these supplies?

A $2.45 C $37.45

B $32.55 D $45.00

Name ______________________________

Practice
Objective
48

DIRECTIONS

Read each question and choose the best answer. Then mark the space for the answer you have chosen.

SAMPLE

A case of 24 cans of fruit juice costs $6.00. What is the cost per can of juice?

A $0.50 C $1.44

B $0.25 D $0.24

1 **A 14-ounce box of breakfast cereal costs $2.66. What is the cost per ounce of the cereal?**

A $0.19 C $0.37

B $0.29 D $0.27

2 **Vicky bought 4 CDs at $12.99 each and 2 CDs at $14.99 each. How much did she have to pay altogether?**

F $51.96 H $21.98

G $29.98 J $81.94

3 **The Harris family bought a computer for $1,590 plus interest. They will make twelve payments of $145.75 each. How much was the interest?**

A $159.00 C $1,749

B $145.75 D $132.50

4 **Mrs. Scanlon paid $84.80 for 32 copies of a paperback novel for her reading class. What was the price for 10 books?**

F $2,7136.00 H $26.50

G $36.50 J $2.65

5 **For the picnic, Jillian bought cases of chips. Each case cost $6.72 and contains 24 individual bags. What was the cost for 12 bags?**

A $0.28 C $3.36

B $0.38 D $4.56

6 **What is the cost of borrowing $1,500 at 9% interest rate for 1 year?**

F $11.25 H $135.00

G $13.50 J $162.00

7 **Ms. Martin borrowed $2,000 at 12% interest per year. If she pays the loan back in two years, how much money must she repay?**

A $240 C $2,240

B $480 D $2,480

Name ____________________

DIRECTIONS

Read each question and choose the best answer. Then mark the space for the answer you have chosen.

SAMPLE

Anna takes two piano lessons each week. What do you need to know in order to find out how much she pays her teacher each week?

A How long the lessons are

B What time the lessons are

C How much each lesson costs

D How much she practices

1 **Larry is saving part of his weekly allowance for new ice skates. The skates cost $55.00 including the sales tax. What do you need to know in order to find out when he will be able to buy the skates?**

A How much the sales tax is

B How much his allowance is

C How many weeks he will save his allowance

D How much he will save each week

2 **It costs $4.25 for each ticket to a movie. Caitlyn has $20. What do you need to know to find out whether she can treat her entire family to a movie?**

F How many tickets are available

G How many family members there are

H When the next show begins

J How much money is needed for the concession stand

3 **Susan sews baby bibs to sell at craft fairs. Each bib takes no more than 10 minutes to make. What do you need to know to find out how long it will take her to sew the bibs for a special order?**

A How many she sold at the last craft fair

B How many she will sew for the order

C How much she charges for each bib

D How many special orders she will have

4 **Tony is saving to buy a new bike lock. He has saved $5.00 each week for 4 weeks. What do you need to know in order to find out when he will have enough money to buy the lock?**

F How much the lock costs

G How much he earns each week

H How much other locks cost

J How much more he could save each week

5 **Max and Cameron are going bowling. They plan to bowl two games at $1.50 each, rent shoes, and have $2.00 for snacks. What do you need to know in order to find out how much the activity will cost?**

A What they will buy for snacks

B How long it will take to bowl two games

C How much they will pay to rent shoes

D What time they will start bowling

Name ______________________________

Practice Objective 50

DIRECTIONS

Read each question and choose the best answer. Then mark the space for the answer you have chosen.

SAMPLE

Margie mixed 3 parts of yellow paint with 2 parts of blue paint. What is the ratio of parts of yellow paint to parts of blue paint?

A 3 to 5 C 2 to 3

B 2 to 5 D 3 to 2

1 **The can of juice concentrate is diluted with 3 cans of water. What is the ratio of cans of concentrate to cans of water?**

A 1 to 3 C 1 to 4

B 3 to 1 D 4 to 1

2 **Katie earned $12 for babysitting 4 hours. What is the ratio of earnings to hours worked?**

F $\frac{12}{4}$ H $\frac{1}{4}$

G $\frac{4}{12}$ J $\frac{1}{3}$

3 **Michael's baseball team won 9 out of 12 games played. What is the ratio of games won to games played?**

A $\frac{9}{3}$ C $\frac{3}{12}$

B $\frac{9}{12}$ D $\frac{12}{9}$

4 **Kim's volleyball team lost 2 games out of 15 played. What is the ratio of losses to wins?**

F 2 to 13 H 15 to 2

G 2 to 15 J 13 to 2

5 **The U.S. flag has 13 stripes and 50 stars. What is the ratio of stars to stripes?**

A 50 to 63 C 50 to 13

B 13 to 50 D 37 to 50

6 **The sixth-grade class has 50 girls and 55 boys. What is the ratio of girls to boys?**

F 50:105 H 55:50

G 105:50 J 50:55

7 **Seven members of the Glee Club are girls. There are 17 members in all. What is the ratio of girls to all the members?**

A 7:10 C 10:7

B 7:17 D 8:17

8 **There are 15 members on Hubbard's softball team. The ratio of girls to boys is 1:2. How many boys are on the team?**

F 5 boys H 8 boys

G 7 boys J 10 boys

Name ______________________________

Practice Objective 50

DIRECTIONS

Read each question and choose the best answer. Then mark the space for the answer you have chosen.

SAMPLE

Alice walks 10 miles per week as part of her exercise program. Which proportion shows how many miles she walks in a year? (1 year = 52 weeks)

A $\frac{10}{1} = \frac{n}{52}$ C $\frac{10}{7} = \frac{52}{n}$

B $\frac{10}{1} = \frac{52}{n}$ D $\frac{7}{10} = \frac{52}{n}$

1 Russell sleeps 8 hours each day. Which proportion shows how many hours he sleeps in a week? (1 week = 7 days)

A $\frac{1}{8} = \frac{n}{7}$ C $\frac{8}{7} = \frac{1}{n}$

B $\frac{8}{1} = \frac{n}{7}$ D $\frac{7}{8} = \frac{1}{n}$

2 If Russell sleeps 8 hours each day, how many hours does he sleep in a week?

F 8 hours H 56 hours

G 54 hours J 48 hours

3 An 8-lb bag of cat food costs $6.40. Which proportion shows the cost per pound?

A $\frac{8}{6.40} = \frac{n}{1}$ C $\frac{6.40}{8} = \frac{n}{1}$

B $\frac{8}{n} = \frac{1}{6.40}$ D $\frac{6.40}{8} = \frac{1}{n}$

4 If an 8-lb bag of cat food costs $6.40, what is the cost per pound?

F $8 H $0.80

G $0.08 J $51.20

5 After exercising, Megan's pulse was 30 beats in 15 seconds. Which proportion shows the number of beats in 1 minute? (1 minute = 60 seconds)

A $\frac{30}{15} = \frac{n}{60}$ C $\frac{30}{60} = \frac{15}{n}$

B $\frac{15}{30} = \frac{n}{60}$ D $\frac{60}{15} = \frac{30}{n}$

6 If Megan's pulse was 30 beats in 15 seconds, how many times did it beat in 1 minute?

F 60 beats H 120 beats

G 90 beats J 150 beats

7 A map uses a scale of 3 cm for every 25 miles. If the map shows a distance of 15 cm, what is the actual distance?

A 5 miles C 75 miles

B $8\frac{1}{4}$ miles D 125 miles

Name ______________________________

Mathematics: Problem Solving

DIRECTIONS

Read each question and choose the best answer. Then mark the space for the answer you have chosen.

SAMPLE

Describe how the next number will change in the following pattern.

2, 4, 6, 8, ...

A It will triple.
B It will double.
C It will increase by 1.
D It will increase by 2.

1 **Claire is saving $7 each week from her paper route earnings to buy new in-line skates. She has saved $28 so far. What do you need to know in order to find out when she will be able to buy the skates?**

A How much her allowance is
B How much the sales tax is
C How much the skates cost
D How much she earns each week

2 **There are 35 green purses, 82 brown purses, 24 red purses and 59 black purses on sale. *About* how many purses are on sale?**

F About 190 purses
G About 200 purses
H About 220 purses
J About 210 purses

3 **Ursala buys 3.9 pounds of hamburger, 1.2 pounds of ground veal, and 1.4 pounds of ground pork. How much ground meat does she buy?**

Estimate. Then decide which answer is most reasonable.

A 6.5 lb　　**C** 5 lb
B 7.9 lb　　**D** 4.6 lb

4 **Lee's birthday is 3 days after Calvin's. Calvin's birthday is the day before Mary's. Lee's birthday is May 21. When is Mary's birthday?**

F May 19　　**H** May 21
G May 20　　**J** May 22

5 **Use your centimeter ruler to measure the perimeter of the figure below.**

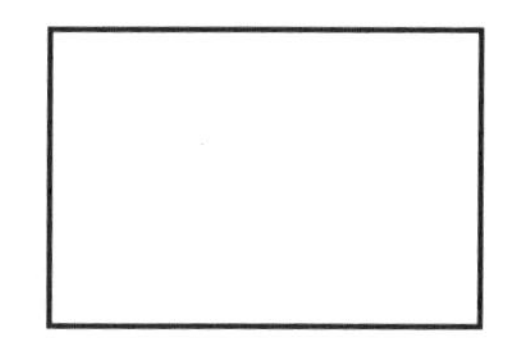

A 9 cm　　**C** 11 cm
B 10 cm　　**D** 12 cm

6 **A fence around the rectangular-shaped tennis courts measures 125 feet by 200 feet. What is the length of fence?**

F 75 ft　　**H** 650 ft
G 325 ft　　**J** 25,000 ft

Name ______________________________

7 **Find the next number in the pattern.**
98, 87, 76, 65, ...

A 44 C 54
B 47 D 56

8 **A school bus traveled from Cayman to Seymour for a baseball game. The map distance between Cayman and Seymour is 3 inches. What is the actual distance between the cities if the scale is 1 in. = $3\frac{1}{2}$ mi?**

F 3 mi H $10\frac{1}{2}$ mi
G $3\frac{1}{2}$ mi J 12 mi

9 **Which pet was the least favorite?**

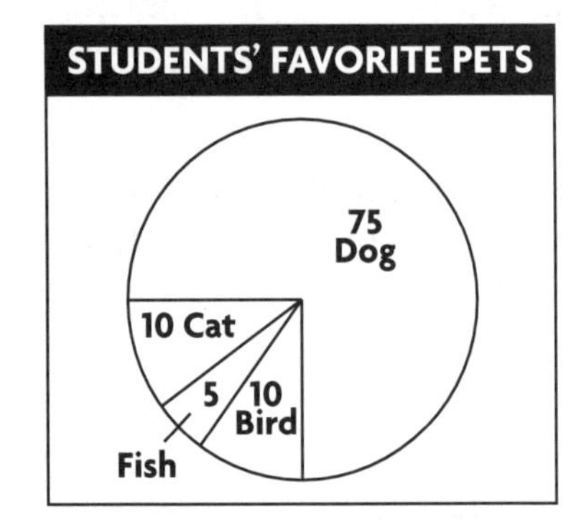

A Dog C Bird
B Cat D Fish

10 **What is the mean (average) number of CDs owned by students in the survey?**

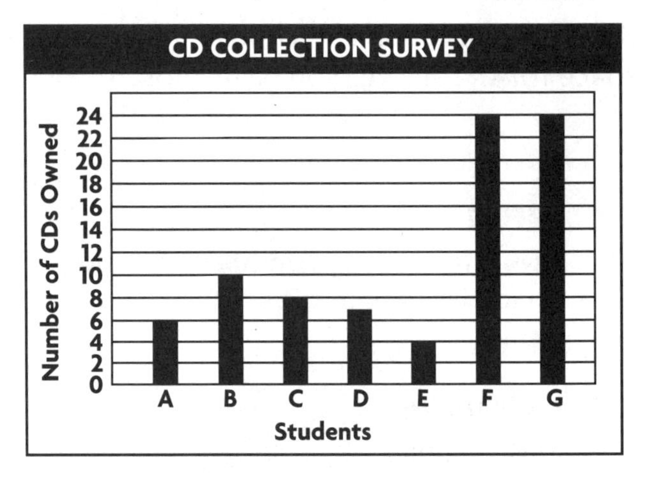

F About 6 CDs H About10 CDs
G About 8 CDs J About 12 CDs

11 **Suppose you toss a nickel and a dime. How many outcomes are possible?**

A 4 outcomes C 2 outcomes
B 3 outcomes D 1 outcome

12 **What is the probability of spinning the letter *A*?**

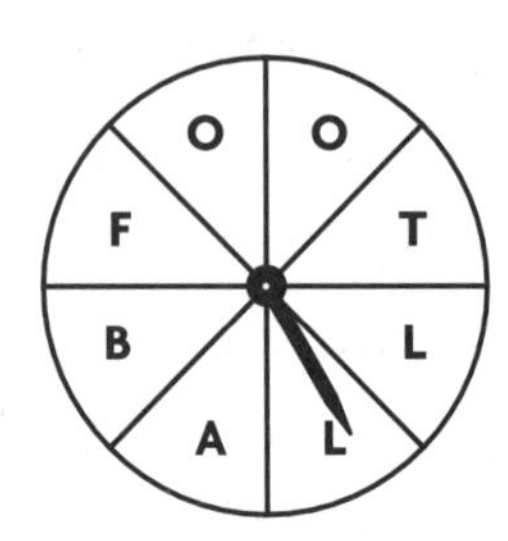

F $\frac{1}{10}$ H $\frac{1}{7}$
G $\frac{1}{8}$ J $\frac{1}{2}$

Name ___________________________

13 **Jenny has a blue blouse and a red blouse. She also has a black skirt and a tan skirt. How many different outfits can she make?**

A 2 outfits
B 3 outfits
C 4 outfits
D 6 outfits

14 **Which measure is most reasonable for the length of a baseball bat?**

F 100 mm
G 100 cm
H 100 m
J 100 km

15 **During a football play, Dack carried the ball 60 feet. How many yards did he run?**

A 5 yd
B 20 yd
C 100 yd
D 180 yd

16 **How many lines of symmetry does this figure have?**

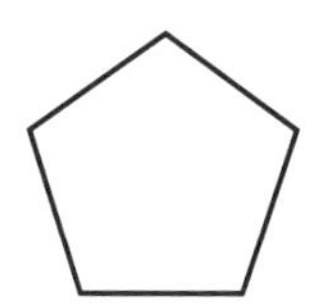

F 1 line
G 2 lines
H 5 lines
J 10 lines

17 **What part of the circle is CA?**

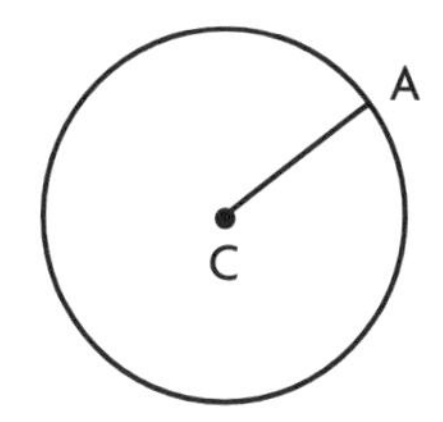

A Diameter
B Radius
C Circumference
D Chord

18 **Which white figure shows where the gray figure would be if the paper were folded along the heavy dark line?**

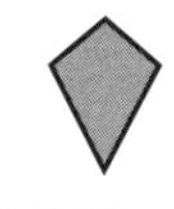

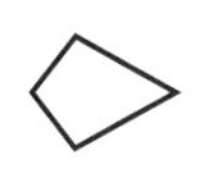

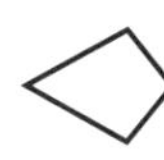

Figure A | Figure B | Figure C | Figure D

F Figure A
G Figure B
H Figure C
J Figure D

19 **What kind of angle is formed by the corner of your math book?**

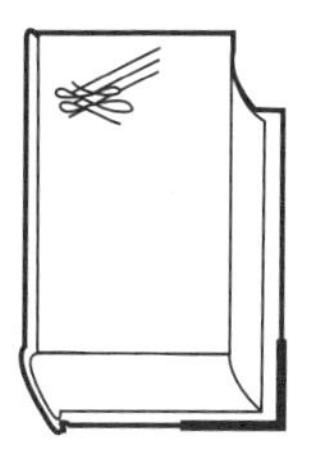

A Obtuse
B Straight
C Right
D Acute

20 **How do you locate the point (3, −4) on the coordinate grid? Begin at 0.**

F Go up 3 spaces, and go left 4 spaces.
G Go down 3 spaces, and go left 4 spaces.
H Go right 3 spaces, and go down 4 spaces.
J Go right 3 spaces, and go up 4 spaces.

21 **A number machine uses a rule to change numbers into other numbers. What number would 8 be changed into?**

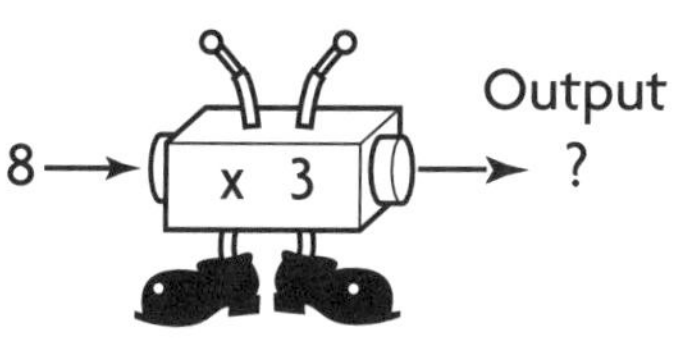

A 32
B 24
C 16
D 8

22 **Which number sentence goes with $35 + 48 = n$?**

F $n + 35 = 48$
G $48 - 35 = n$
H $n \times 35 = 48$
J $n - 48 = 35$

23 **What is the value of 8 in the number 3,805,496?**

A 8 millions
B 8 hundred thousands
C 8 ten thousands
D 8 thousands

24 **Which fraction is less than $\frac{5}{8}$?**

F $\frac{1}{2}$
G $\frac{3}{4}$
H $\frac{5}{6}$
J $\frac{9}{10}$

25 **What is the least common multiple (LCM) of 4 and 6?**

A 4
B 6
C 12
D 24

26 **What decimal shows the part of the whole square that is shaded?**

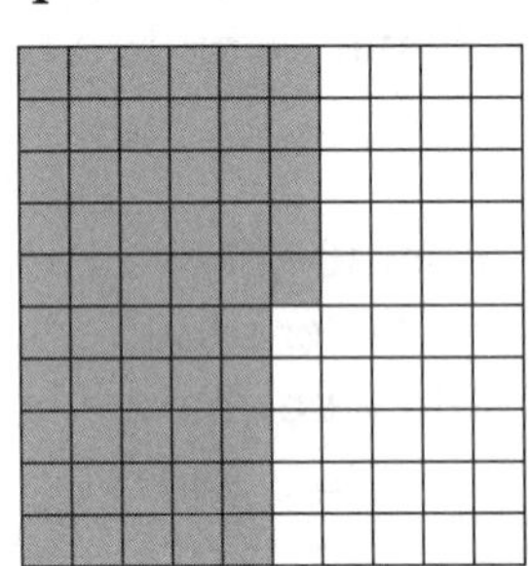

F 0.45
G 0.55
H 0.65
J 0.75

Mathematics: Procedures

DIRECTIONS

Read each question and choose the best answer. Then mark the space for the answer you have chosen. If a correct answer is *not here,* mark the space for NH.

SAMPLE

$$1.5 \times 3$$

A 45
B 4.5
C 0.045
D 0.45
E NH

1

$$7\frac{3}{5} + 4\frac{1}{5}$$

A $\frac{4}{5}$
B $\frac{1}{2}$
C $11\frac{2}{5}$
D $11\frac{1}{10}$
E NH

2

$$\frac{1}{8} + \frac{1}{4}$$

F $\frac{1}{32}$
G $\frac{1}{12}$
H $\frac{1}{2}$
J $\frac{3}{8}$
K NH

3

$$5\frac{3}{4} - 2\frac{1}{4}$$

A $1\frac{1}{2}$
B $3\frac{1}{4}$
C $3\frac{1}{2}$
D 8
E NH

Name ________________________________

4 $\begin{array}{r} \frac{7}{10} \\ -\ \frac{3}{5} \\ \hline \end{array}$

F $\frac{4}{5}$ H $\frac{4}{15}$ K NH
G $\frac{3}{5}$ J $\frac{1}{10}$

5 $\frac{2}{3} \times \frac{4}{7} =$

A $\frac{8}{7}$ C $\frac{8}{21}$ E NH
B $\frac{4}{21}$ D $\frac{6}{10}$

6 $\frac{1}{4} \div \frac{1}{8} =$

F $\frac{1}{32}$ H 2 K NH
G $\frac{1}{2}$ J 4

7 $\begin{array}{r} 28.32 \\ -\ 13.54 \\ \hline \end{array}$

A 4.78 C 15.22 E NH
B 14.88 D 15.78

8 5.3 + 8.19 =

F 8.72 H 13.49 K NH
G 13.22 J 87.2

9 $\begin{array}{r} 2.58 \\ \times\ 1.3 \\ \hline \end{array}$

A 0.3354 C 33.54 E NH
B 3.354 D 335.4

10 **The population of Wisconsin in 1990 was 4,891,769 people. What is that number rounded to the nearer million?**

F 3,000,000 J 6,000,000
G 4,000,000 K NH
H 5,000,000

11 **Jerry buys a 9-foot piece of ribbon to use for wrapping presents. How many $\frac{3}{4}$-feet sections can he cut?**

A 12 sections D $6\frac{3}{4}$ sections
B $9\frac{3}{4}$ sections E NH
C $8\frac{1}{4}$ sections

12 **Kyle buys a blue shirt for $19.59, a green shirt for $12.99, and a white shirt for $21.98. What was the total cost before tax was added?**

F $54.56 H $41.54 K NH
G $42.36 J $32.58

13 **Corrine buys a CD for $12.95 and a book for $8.95. She pays for them with a $20 bill and a $5 bill. How much change should she receive?**

A $3.10 C $15.00 E NH
B $12.05 D $21.90

14 **A single can of dog food costs $0.85. A package of 5 cans costs $3.50. How much can be saved on 5 cans by buying the package?**

F $0.55 H $2.65 K NH
G $ 0.85 J $4.25

15 **There were 150 boxes to load in a truck. Dan has loaded 75 of them. What percent of the boxes has been loaded?**

A 2% C 25% E NH
B 5% D 50%

Name ______________________________

Mathematics: Problem Solving

DIRECTIONS

Read each question and choose the best answer. Then mark the space for the answer you have chosen.

SAMPLE

How will the next number change in the following pattern?

2, 4, 8, 16, ...

A It will triple.
B It will double.
C It will increase by 1.
D It will increase by 2.

1 **Tyler is saving to buy a new bike. He has saved $10 each week for 9 weeks from his baby-sitting jobs. What do you need to know in order to find out when he will be able to buy the bike?**

A How much the bike costs
B How much he has saved so far
C How many baby-sitting jobs he has each week
D How much he spends each week

2 **On Monday 275 students visited the museum. On Tuesday 362 students visited the museum, and on Wednesday 295 students visited. About how many students visited the museum in all?**

F About 500 students
G About 800 students
H About 850 students
J About 1,000 students

3 **Jonathon spends $62.98 for a jacket, $32.45 for a pair of pants, and $19.35 for a shirt. How much does he spend all together?**

Estimate. Then decide which answer is most reasonable.

A $114.78 **C** $82.30
B $95.46 **D** $51.80

4 **Each day Dennis practices playing his drums. On Monday he practiced 15 minutes, on Tuesday 20 minutes, and on Wednesday 25 minutes. If he continues at the same rate, on which day will he practice for 45 minutes?**

F Friday **H** Sunday
G Saturday **J** Monday

5 **Chicago is about 300 miles from St. Louis. How long will it take an auto traveling at 60 miles per hour to go from Chicago to St. Louis?**

A 4 hours **C** 5 hours
B 4.5 hours **D** 6 hours

6 **Shae's driveway is 12 feet wide and 50 feet long. What is the area of the driveway?**

F 62 ft^2 **H** 600 ft^2
G 124 ft^2 **J** 3,100 ft^2

7 **Find the next term in the number pattern.**

5, 25, 125, ...

A 3,125 **C** 325
B 625 **D** 225

Name ______________________________

8 **The Altemeyer family traveled from Maryville to Louisville. The distance on the map is $6\frac{1}{4}$ inches. What is the actual distance if the scale is 1 in. = 100 miles?**

F 600 miles
G 625 miles
H 650 miles
J 675 miles

9 **What has been the trend in bicycle sales?**

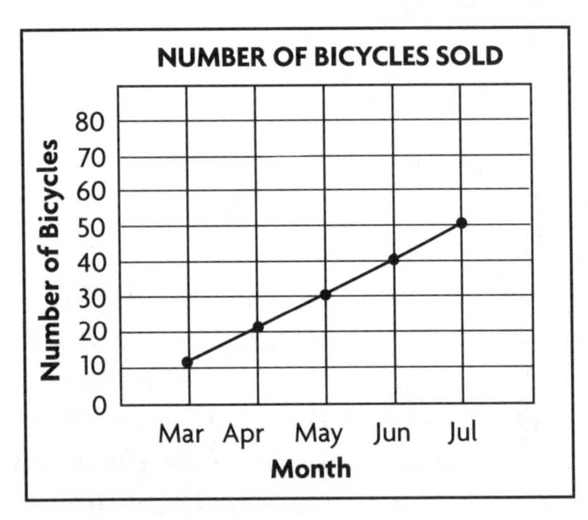

A Increasing each month
B Decreasing each month
C Staying the same
D Mostly increasing each month

10 **Which measure of central tendency best represents this data?**

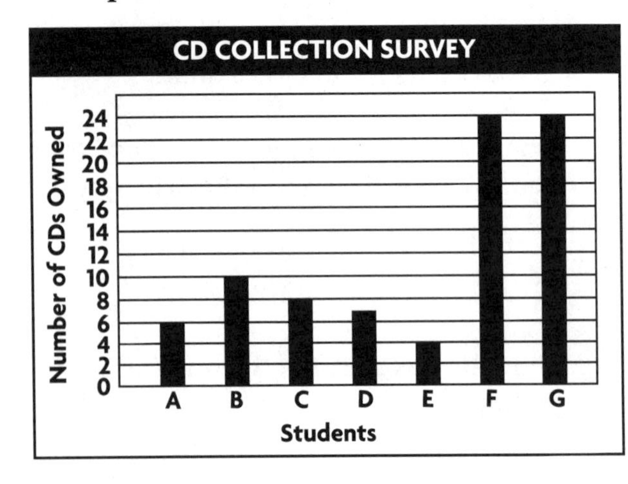

F Mean
G Mode
H Median
J Range

11 **Suppose you toss a coin and a number cube. How many outcomes are possible?**

A 2 outcomes
B 4 outcomes
C 8 outcomes
D 12 outcomes

12 **What is the probability of a game piece landing on a number divisible by 4?**

1	2	3	4
5	6	7	8
9	10	11	12
13	14	15	16

F $\frac{1}{16}$
G $\frac{3}{16}$
H $\frac{1}{4}$
J $\frac{1}{2}$

13 **Deidra has three wigs and four pairs of dark glasses. How many disguises can she create using the wigs and glasses?**

A 6 disguises
B 7 disguises
C 12 disguises
D 14 disguises

14 **Which measure is the most reasonable for the capacity of a kitchen sink?**

F 5 mL
G 50 mL
H 900 mL
J 54 L

15 **A bottle of water holds 16 fluid ounces. How many cups does it hold?**

A 1 cup
B $1\frac{1}{2}$ cups
C 2 cups
D $2\frac{1}{2}$ cups

Name ______________________________

16 **How many lines of symmetry does this figure have?**

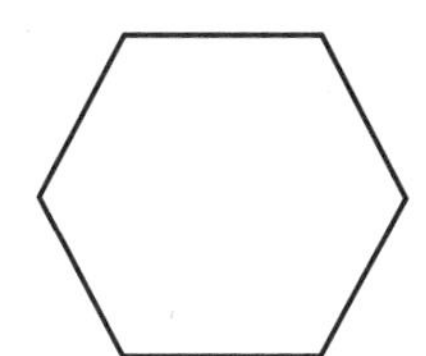

F 12 lines
G 8 lines
H 6 lines
J 3 lines

17 **What part of the circle is *AB*?**

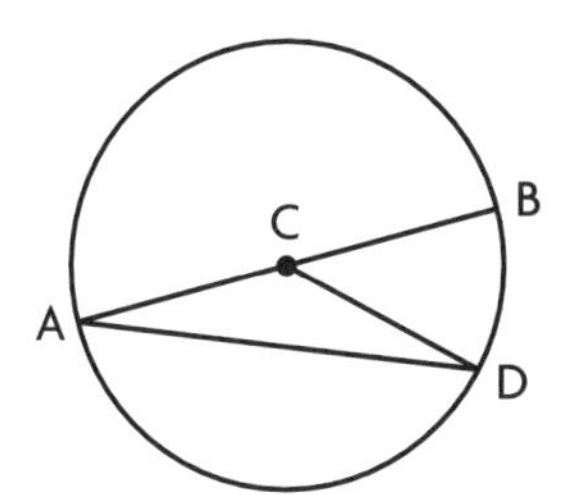

A Diameter
B Radius
C Circumference
D Pi

18 **What pattern of transformations was used to move the figure from position 1 (through positions 2 and 3) to position 4?**

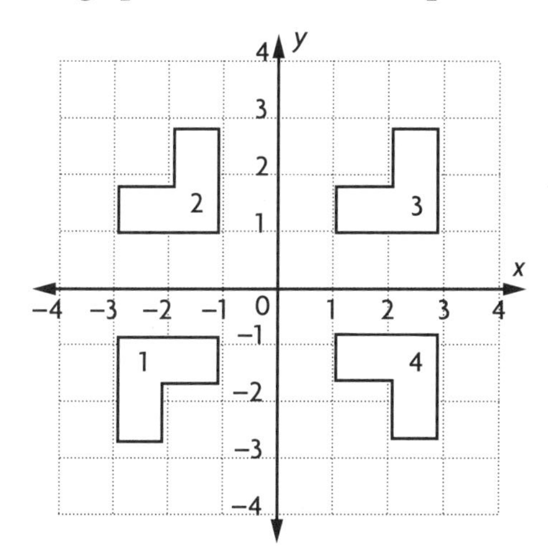

F Rotation, reflection, rotation
G Reflection, translation, reflection
H Translation, reflection, translation
J Rotation, translation, reflection

19 **What kind of angle is formed by the compass opening?**

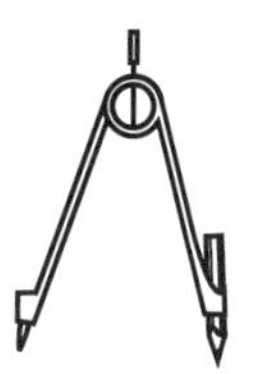

A Obtuse
B Straight
C Right
D Acute

20 **Name the ordered pair for point *E*?**

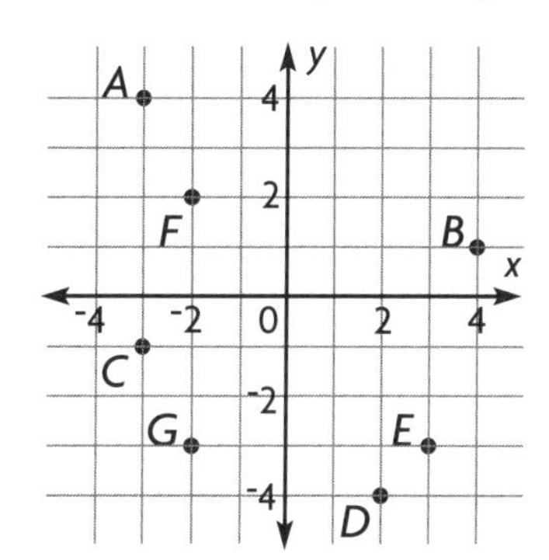

F (3, 3)
G (−3, 3)
H (−3, −3)
J (3, −3)

21 **If the input is 8, what is the output?**

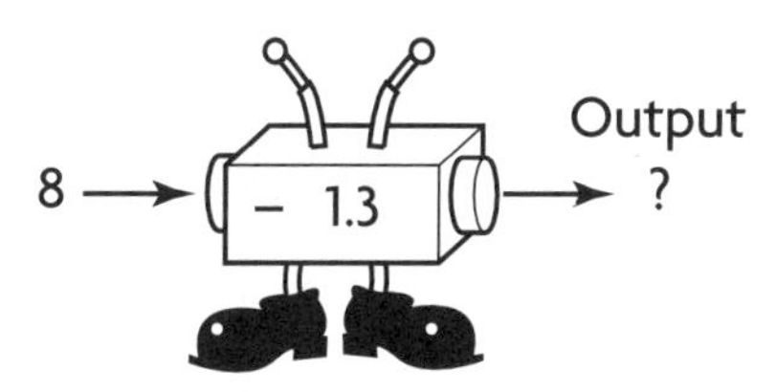

A −6.7
B 5.7
C 6.7
D 9.3

22 **If $38 \times 47 = n \times 38$, then $n =$**

F 9
G 85
H 47
J 1,786

23 Which is the standard form of sixty-one thousand, three hundred nine and eighty-five thousandths?

A 61,390.85
B 61,390.085
C 61,309.85
D 61,309.085

24 Which shows the mixed numbers written from greatest to least?

F $14\frac{2}{3}, 13\frac{7}{8}, 14\frac{4}{9}, 14\frac{3}{10}$
G $13\frac{7}{8}, 14\frac{4}{9}, 14\frac{3}{10}, 14\frac{2}{3}$
H $14\frac{2}{3}, 14\frac{4}{9}, 14\frac{3}{10}, 13\frac{7}{8}$
J $14\frac{4}{9}, 14\frac{2}{3}, 14\frac{3}{10}, 13\frac{7}{8}$

25 What is the least common multiple (LCM) of 12 and 8?

A 8　　C 24
B 12　　D 48

26 What fraction of the whole square is shaded?

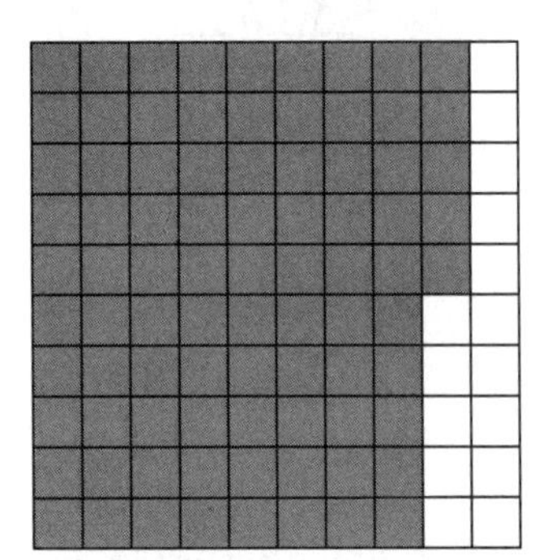

F $8\frac{1}{2}$　　H $\frac{17}{20}$
G $\frac{85}{90}$　　J $\frac{4}{5}$

Mathematics: Procedures

DIRECTIONS

Read each question and choose the best answer. Then mark the space for the answer you have chosen. If a correct answer is *not here*, mark the space for NH.

SAMPLE

$$\begin{array}{r} 1.9 \\ \times\ 1.5 \\ \hline \end{array}$$

A 285　　C 2.85　　E NH
B 28.5　　D 0.285

1

$$\begin{array}{r} 5\frac{1}{6} \\ +\ 3\frac{1}{6} \\ \hline \end{array}$$

A $8\frac{1}{12}$　　C $8\frac{1}{3}$　　E NH
B $8\frac{1}{4}$　　D $8\frac{2}{3}$

2

$$\begin{array}{r} \frac{1}{4} \\ +\ \frac{1}{6} \\ \hline \end{array}$$

F $\frac{1}{24}$　　H $\frac{1}{10}$　　K NH
G $\frac{5}{12}$　　J $\frac{1}{48}$

3

$$\begin{array}{r} 6\frac{7}{8} \\ -\ 2\frac{3}{4} \\ \hline \end{array}$$

A $4\frac{5}{8}$　　C $4\frac{1}{4}$　　E NH
B $4\frac{1}{2}$　　D $4\frac{3}{4}$

Name ______________________________

4 $\begin{array}{r} \frac{11}{12} \\ -\ \frac{3}{4} \\ \hline \end{array}$

F $\frac{1}{8}$ H $\frac{1}{2}$ K NH

G $\frac{1}{6}$ J 1

5 $\frac{1}{5} \times \frac{5}{8} =$

A $\frac{1}{2}$ C $\frac{6}{13}$ E NH

B $\frac{3}{21}$ D $\frac{1}{8}$

6 $\frac{3}{5} \div \frac{1}{10} =$

F $\frac{3}{50}$ H 1 K NH

G $\frac{1}{3}$ J 2

7 $\begin{array}{r} 72.45 \\ -\ 34.97 \\ \hline \end{array}$

A 37.48 C 38.58 E NH

B 42.52 D 107.42

8 $3.45 - 0.126 =$

F 3.324 H 3.326 K NH

G 3.334 J 3.576

9 $\begin{array}{r} 3.09 \\ \times\ 0.38 \\ \hline \end{array}$

A 1.1742 C 117.42 E NH

B 11.742 D 1,174.2

10 **The population of Florida in 1990 was 12,938,071. What is the number rounded to the nearer million?**

F 10,000,000 J 13,000,000

G 12,000,000 K NH

H 12,900,000

11 **Candace buys $\frac{3}{4}$ of a package of 24 items that is on sale for $3.28. How much did she spend?**

A $2.46 C $32 E NH

B $18 D $78.72

12 **Corey buys a model airplane for $7.38, a model car for $8.49, and two model trucks for $12.58 each. What was the total cost before tax?**

F $41.13 H $28.45 K NH

G $30.02 J $27.25

13 **Dixie buys a magazine for $3.95 including tax, two books for $14.59 each including tax, and a greeting card for $0.95 including tax. She pays for the items with a $50 bill. How much change should she receive?**

A $34.08 C $24.08 E NH

B $25.92 D $15.92

14 **A single package of chips costs $1.29. A carton of 10 packages costs $10.60. How much can be saved on 20 packages by buying the carton?**

F $2.30 H $4.60 K NH

G $2.90 J $9.31

15 **Casey buys a blouse on sale at 25% off the original price of $24.92. How much money did she save?**

A $6.23 C $18.69 E NH

B $12.46 D $31.15

Name ______________________

Mathematics: Problem Solving

DIRECTIONS

Read each question and choose the best answer. Then mark the space for the answer you have chosen.

SAMPLE

How will the next number change in the following pattern?

3, 9, 27, 81, ...

A It will triple.
B It will double.
C It will increase by 3.
D It will increase by 6.

1 **Jessica is going to the movies with her two sisters. Admission is $4.50 each and they usually buy snacks at the concession stand. What do you need to know in order to find out how much it will cost them to go to the movies and buy snacks?**

A How much admission tickets will cost
B How many people will be in the group
C How much each person usually spends at the concession stand
D How long the movie lasts

2 **The movie theater can seat 289 people. It was sold out for the last 19 shows. About how many people saw the last 19 shows?**

F About 4,000 people
G About 6,000 people
H About 8,000 people
J About 10,000 people

3 **Eric buys two computer games for $49.99 each and an educational piece of software for $40.02. The sales tax is 7%. How much money does he give the clerk?**

A $149.80 C $140
B $148.40 D $9.80

4 **The Witt family plans to drive 300 miles the first day on their vacation and to drive a total of 500 miles by the end of the second day. If y = miles they drive the second day, which equation would you use to find how many miles they drive the second day?**

F $500 + 300 = y$
G $y - 300 = 500$
H $300 - y = 500$
J $300 + y = 500$

5 **If the length of each side of a cube is doubled, how does its volume change?**

A It is 2 times the original cube.
B It is 4 times the original cube.
C It is 6 times the original cube.
D It is 8 times the original cube.

6 **The floor measures 30 feet by 25 feet. Each tile measures 1 square foot and costs $2.50. What will it cost to cover the floor?**

F $137.50 H $750.00
G $275.00 J $1,875.00

Name ________________________________

7 **How many cubes would there be in the next figure in this sequence?**

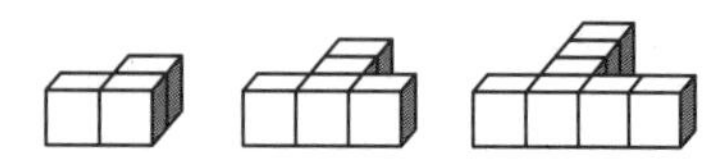

A 18 cubes
B 16 cubes
C 10 cubes
D 9 cubes

8 **The map distance of the bike trail is 3 inches from one parking area to the other parking area. What is the roundtrip distance if 1 in. = 2 miles?**

F 6 miles
G 9 miles
H 12 miles
J 24 miles

9 **Sonya is using this data to make a circle graph. What percent of the circle will the salad section be?**

Favorite Lunch	
Pizza	30 votes
Chicken	25 votes
Salad	20 votes
Sandwich	25 votes

A 50%
B 30%
C 25%
D 20%

10 **The table shows the scores students received on their last math quiz. What is the median score?**

Scores on Math Quiz				
58	100	64	92	76
80	82	98	98	97
98	88	96	87	87

F 82
G 87
H 88
J 98

11 **Suppose you spin the spinner once and toss a number cube labeled from 1 to 6 once. How many outcomes are possible?**

A 10 outcomes
B 12 outcomes
C 18 outcomes
D 24 outcomes

12 **A stack of playing cards is numbered from 1 to 10. What is the probability that Shawna will *not* choose a card numbered 2 or 6?**

F 10%
G 20%
H 75%
J 80%

13 **Doris, Tince, Kay, and Elvin line up to get their school pictures taken. If Doris is always first, how many ways can they line up?**

A 3 ways
B 6 ways
C 8 ways
D 16 ways

14 **Dudley and three friends measured the distance from home plate to the pitcher's mound. Which measurement was more precise?**

F 14 yd
G $13\frac{1}{3}$ yd
H 13 yd
J 40 ft

15 **A can of coffee weighs 2 lb 7 oz. How many ounces will 3 cans weigh?**

A 27 oz
B 39 oz
C 78 oz
D 117 oz

16 **How many lines of symmetry does this figure have?**

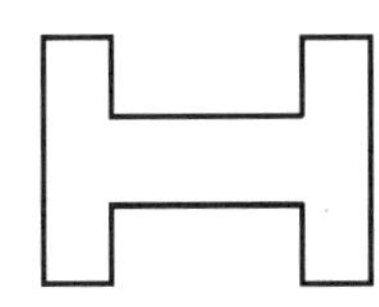

F 0 lines
G 1 line
H 2 lines
J 8 lines

17 **What part of the circle is *AB*?**

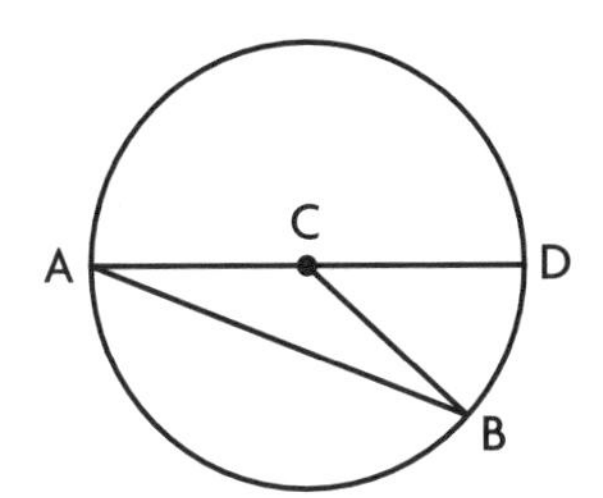

A Chord
B Radius
C Circumference
D Diameter

18 **If you translate triangle *ABC* 4 units to the left, what are the coordinates of the new triangle?**

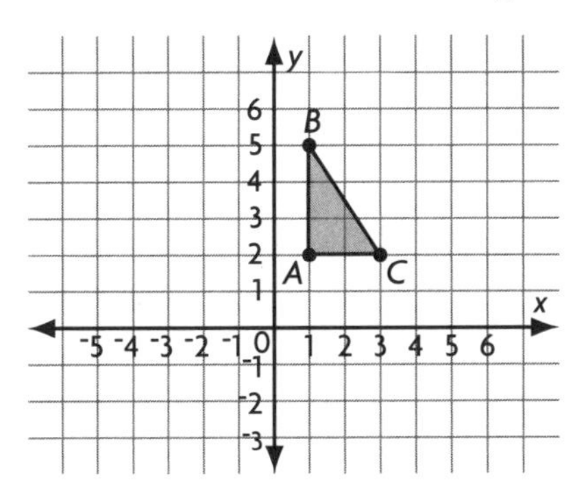

F $A'(^-3, 2)$, $B'(^-3, 5)$, $C'(^-1, 2)$
G $A'(3, 2)$, $B'(3, 5)$, $C'(1, 2)$
H $A'(1, 0)$, $B'(1, 3)$, $C'(3, 0)$
J $A'(^-4, 2)$, $B'(^-4, 5)$, $C'(0, 2)$

19 **What kind of angle is formed between the book covers?**

A Obtuse
B Straight
C Right
D Acute

20 **Name the ordered pair for point *G*.**

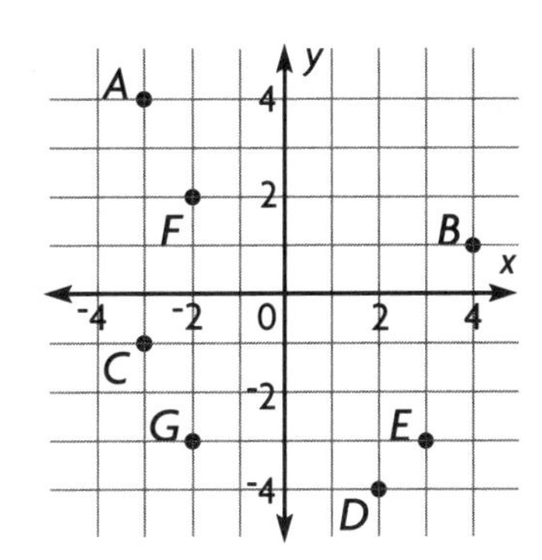

F (2, 3)
G $(^-2, 3)$
H $(^-2, ^-3)$
J $(2, ^-3)$

21 **If the input is 48, what is the output?**

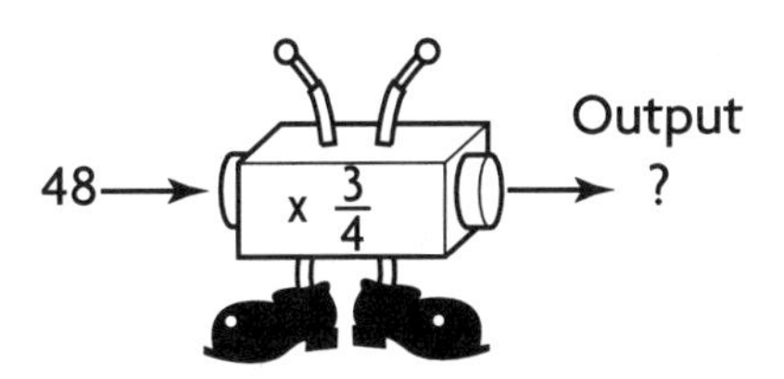

A 12
B 36
C 42
D 64

22 **If $3 \times 98 = 3 \times (n + 8)$, then $n =$**

F 24
G 90
H 270
J 294

Name ______________________________

23 **Which shows the numbers written from least to greatest?**

A 35.79, 37.59, 39.57, 35.97
B 39.57, 37.59, 35.97, 35.79
C 39.57, 37.59, 35.79, 35.97
D 35.79, 35.97, 37.59, 39.57

24 **What is $8\frac{7}{12}$ written as a fraction?**

F $\frac{103}{12}$ **H** $\frac{58}{12}$
G $\frac{103}{8}$ **J** $\frac{96}{7}$

25 **What is the greatest common factor (GCF) of 32 and 48?**

A 48 **C** 16
B 32 **D** 8

26 **What percent of the whole square is shaded?**

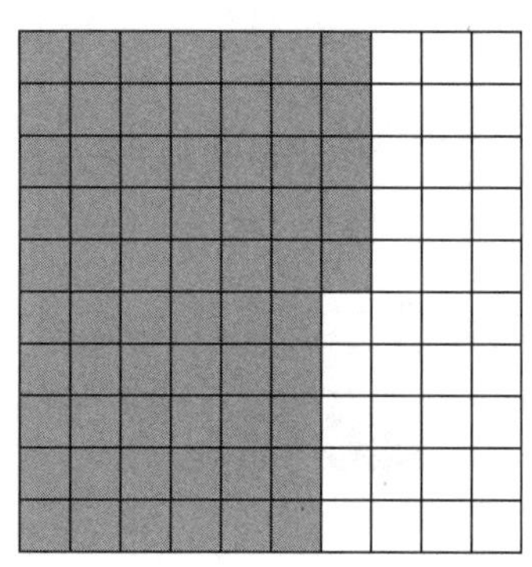

F 13% **H** 65%
G 56% **J** 85%

Mathematics: Procedures

DIRECTIONS

Read each question and choose the best answer. Then mark the space for the answer you have chosen. If a correct answer is *not here,* mark the space for NH.

SAMPLE

$$2.08 \times 3.5$$

A 72.8 **C** 7.28 **E** NH
B 72.08 **D** 7.028

1 $$21\frac{3}{4} + 15\frac{2}{3}$$

A $36\frac{1}{7}$ **C** $37\frac{11}{24}$ **E** NH
B $36\frac{5}{7}$ **D** $37\frac{5}{12}$

2 $$\frac{5}{6} + \frac{3}{8}$$

F $\frac{1}{7}$ **H** $1\frac{11}{48}$ **K** NH
G $\frac{4}{7}$ **J** $1\frac{5}{24}$

3 $$5\frac{1}{3} - \frac{3}{8}$$

A $3\frac{1}{3}$ **C** $4\frac{1}{3}$ **E** NH
B $3\frac{1}{2}$ **D** $4\frac{1}{2}$

4 $\begin{array}{r} \frac{3}{5} \\ -\frac{1}{3} \\ \hline \end{array}$

F $\frac{2}{15}$ H $\frac{2}{3}$ K NH

G $\frac{4}{15}$ J $\frac{14}{15}$

5 $\frac{7}{12} \times \frac{4}{7} =$

A $\frac{1}{4}$ C $\frac{10}{19}$ E NH

B $\frac{21}{108}$ D $\frac{4}{5}$

6 $\frac{7}{8} \div \frac{3}{4} =$

F $\frac{6}{7}$ H $1\frac{1}{6}$ K NH

G $\frac{21}{32}$ J $1\frac{11}{21}$

7 $\begin{array}{r} 23.009 \\ -\ 12.345 \\ \hline \end{array}$

A 10.664 C 11.664 E NH

B 11.344 D 11.764

8 $0.63 \times 1.8 =$

F 1.134 H 10.74 K NH

G 11.34 J 1.074

9 $\begin{array}{r} 3.14 \\ \times\ 2.05 \\ \hline \end{array}$

A 0.6437 C 6.437 E NH

B 0.885 D 8.85

10 **The population of California in 1990 was 29,758,213. What is the number rounded to the nearer million?**

F 20,000,000 J 30,000,000

G 25,000,000 K NH

H 29,000,000

11 **Manuel buys a carton of 12 sodas for \$3.98. He keeps $\frac{1}{3}$ of the sodas and divides the rest equally among 4 friends. How many sodas will each friend receive?**

A 2 sodas C 4 sodas E NH

B 3 sodas D 6 sodas

12 **Bonita buys 4 notebooks for \$4.90 each and 8 pads of paper for \$2.00 each. Sales tax is 5%. How much did she pay in all?**

F \$0.98 H \$35.60 K NH

G \$1.78 J \$37.38

13 **Xavier uses a \$50 bill to buy 2 model cars for \$9.90 each, 2 tubes of paint for \$3.50 each, and 2 tubes of glue for \$2.00 each. Tax is 5%. How much money does he have left?**

A \$1.54 C \$30.80 E NH

B \$17.46 D \$32.34

14 **April can buy a 24-ounce box of cereal for \$4.32. A 12-ounce box costs \$2.52. How much more per ounce does the cereal in the smaller box cost than the cereal in the larger box?**

F \$1.70 H \$0.18 K NH

G \$0.21 J \$0.03

15 **What is the cost of borrowing \$1,800 at 8% interest for 2 years?**

A \$144 C \$288 E NH

B \$200 D \$360

Correlation of the
Sixth Grade
Virginia Mathematics Standards of Learning
to *Math Advantage* © 1999

CRITERIA	LESSON/APPLICATION PAGE REFERENCES
Number and Number Sense	
6.1 The student will identify representations of a given percent and describe orally and in writing the equivalent relationship between fractions, decimals, and percents.	22–25, 32, 33, 53, 55, 78, 80, 81, 151, 157, 336–339, 340–341, 345, 346, 347, 353, 355, 361, 397, 398, 403, 419, 461, 463, 464–465, 467, 468–469, 470–473, 478, 489, 495, 514, 516 H 19, 28, 29, 35, 42, 67, 77 C 66–67, 92–93 P 3, 76, 77, 107, 108 PS 3, 4, 5, 6, 9, 11, 76, 77, 80, 108, 109, 110, 111, 112, 114, 115, 116, 117, 118 R 3, 76, 77, 107, 108 E-LAB 24
6.2 The student will describe and compare two sets of data using ratios and will use appropriate notations such as a/b, a to b, and a:b.	265, 330D, 332–333, 346, 366D H 27, 66 C 70–71 P 74 PS 74, 76, 77, 82, 85, 86, 87, 88, 91 R 74 E–LAB 17
6.3 The student will explain orally and in writing the concepts of prime and composite numbers.	84–85, 98, 152, 154 H 2, 46 C 36–37 P 15 PS 15, 16, 17 R 15
6.4 The student will compare and order whole numbers, fractions, and decimals, using concrete materials, drawings or pictures, and mathematical symbols.	18–21, 22–25, 32, 33, 76, 81, 91, 99, 157, 233, 267, 339, 355, 443, 463, 476–477, 478, 504, 518, 551 H 7, 11, 15, 16, 42, 78 C 2–3, 8–9, 12–13, 78–79 P 2, 3, 110 PS 2, 3, 4, 5, 8, 9, 11, 13, 14 R 2, 110
6.5 The student will identify and represent integers on a number line.	30, 33, 75, 76, 466 C 78–79, 80–81, 82

CRITERIA	LESSON/APPLICATION PAGE REFERENCES
Computation and Estimation	
6.6 The student will	
solve problems that involve addition, subtraction and/or multiplication with fractions and mixed numbers, with and without regrouping, that include like and unlike denominators of 12 or less and express their answers in simplest form; and	100C, 102–103, 104–105, 106–107, 108, 109, 110–111, 116, 117, 118D, 119, 120–121, 122–123, 124–125, 126–129, 131, 132, 133, 134D, 136–139, 140–141, 142–143, 147, 150, 151, 155, 156, 157, 163, 175, 193, 211, 217, 233, 249, 381, 403, 419, 461, 519 H 17, 18, 19, 34, 48, 49, 50, 51, 52 C 40–41 P 20, 21, 22, 23, 25, 26, 27, 29, 30, 31 PS 20, 21, 22, 23, 25, 26, 27, 29, 30, 31 R 20, 21, 22, 23, 25, 26, 27, 29, 30, 31 E-LAB 6
find the quotient, given a dividend expressed as a decimal through thousandths and a divisor expressed as a decimal to thousandths with exactly one non-zero digit. For divisors with more than one non-zero digit, estimation and calculators will be used.	68–69, 70–73, 71, 72, 74, 76, 79, 80, 81, 133 H 13, 46 C 6–7 P 14 E-LAB 3
6.7 The student will use estimation strategies to solve multistep practical problems involving whole numbers, decimals, and fractions.	50–53, 54, 55, 58–61, 64–67, 74, 79, 112–115, 114, 116, 117, 129, 130–131, 132, 133, 138, 151, 157, 169, 193, 211, 249, 267, 283, 289, 305, 321, 323, 329, 437 H 10, 45, 46, 49, 51 C 6–7 P 10, 11, 13, 24, 28 PS 1, 3, 4, 5, 6, 8, 9, 10, 12, 14, 16, 18, 19, 21, 22, 23, 24, 25, 27, 28, 39, 41, 48, 50, 51, 52, 54, 56, 59, 67, 69, 70, 74, 80, 82, 95, 96, 99, 101, 102, 103, 105, 106, 108, 111, 113, 114, 115, 118, 119, 124, 125, 126, 127 R 10, 24, 28
6.8 The student will solve multistep consumer application problems involving fractions and decimals and present data and conclusions in paragraphs, tables, or graphs.	32, 42, 47, 52, 53, 56, 60, 61, 62, 63, 66, 67, 69, 73, 74, 75, 80, 88, 91, 95, 133, 135, 147, 149, 156, 175, 194, 209, 218D, 220, 239, 244, 250D, 251, 256, 260, 261, 275, 276, 293, 297, 308, 314, 334–335, 338, 348D, 354–355, 356–357, 358–359, 360–361, 364, 375, 379, 398, 400, 403, 407, 419, 424, 446, 449, 453, 473, 475, 479, 489, 495, 503, 505, 527, 539 H 36, 68, 69 C 38–39, 94–95 P 33, 75, 80, 82 PS 13, 23, 33, 75, 80, 81, 82, 83, 85, 89, 120, 122 R 33, 75, 82

CRITERIA	LESSON/APPLICATION PAGE REFERENCES
Measurement	
6.9 The student will compare and convert units of measures for length, weight/mass, and volume within the U.S. Customary system and within the metric system and estimate conversions between units in each system:	
length--part of an inch (1/2, 1/4, and 1/8), inches, feet, yards, miles, millimeters, centimeters, meters, kilometers;	47, 119, 123, 195, 383, 406–407, 408–409, 410–411, 416, 418, 419, 423, 429, 458 H 24, 72 C 72–73 P 92, 93, 94 PS 92, 93, 94, 95, 99, 102, 105 R 92, 93, 94
weight/mass--ounces, pounds, tons, grams, and kilograms;	406–407, 408–409, 418, 423, 429, 458, 495 H 24, 72 C 72–73 P 92, 93 PS 92, 94, 95, 99 R 92, 93
liquid volume--cups, pints, quarts, gallons, milliliters, and liters; and	406–407, 408–409, 418, 423, 458, 535 H 24, 72 C 72–73 P 92, 93 PS 92, 93, 96, 97, 99 R 92, 93
area--square units.	25, 273–275, 420D, 424–425, 436, 443, 455, 456, 457, 460 H 74 C 76–77 P 98 PS 99, 100, 101, 102, 103
6.10 The student will estimate and then determine length, weight/ mass, area, and liquid volume/ capacity, using standard and nonstandard units of measure.	25, 101, 119, 123, 195, 273–275, 383, 405, 410–411, 416, 420D, 421, 422–423, 424–425, 426–429, 436, 437, 443, 447, 455, 456, 457, 458, 460, 461 H 74 C 76–77 P 94, 97, 98, 99 PS 97, 99, 100, 101, 102, 103, 104, 105 R 94, 97, 98
6.11 The student will determine if a problem situation involving polygons of four sides or less represents the application of perimeter or area and apply the appropriate formula.	149, 345, 415, 424–425, 429, 431, 436, 455, 459, 460 H 74 PS 98

CRITERIA	LESSON/APPLICATION PAGE REFERENCES
6.12 The student will create and solve problems by finding the circumference and/or area of a circle when given the diameter or radius. Using concrete materials or computer models, the student will derive approximations for pi from measurements for circumference.	416–417, 432–433, 434–435, 436, 437, 447, 459, 461 H 37, 75 C 74–75, 102–103 P 101 PS 101, 102, 104, 105 R 101 E-LAB 21, 22
6.13 The student will estimate angle measures using 45°, 90°, and 180° as referents and use the appropriate tools to measure the given angles.	158D, 163, 164–165, 174, 175, 187, 212, 214, 217, 305, 355, 461 H 53, 54 P 36, 37 PS 35, 36, 37 R 36
Geometry	
6.14 The student will identify, classify, and describe the characteristics of plane figures including similarities and differences.	169, 172–173, 174, 175, 199, 203, 211, 215, 216, 217, 407, 465, 519 H 21, 22, 54 C 16–17, 20–21 P 38 PS 38, 39, 40, 41, 46 R 38
6.15 The student will determine congruence of segments, angles, and polygons by direct comparison, given their attributes. Examples of noncongruent and congruent figures will be included.	166–169, 174, 175, 178–181, 370–371, 380, 398 H 23, 54, 69 C 18–19 P 37, 84 PS 37, 84, 88 R 37, 84
6.16 The student will construct the perpendicular bisector of a line segment and an angle bisector, using a compass and straightedge.	170–171 C 44–45
6.17 The student will sketch, construct models, and classify rectangular prisms, cones, cylinders, and pyramids.	194D, 195, 196–199, 200–201, 202–203, 208–209, 210, 211, 215, 216, 217, 221, 231, 233, 249, 283, 305, 323, 329, 365, 381 H 22, 23, 56, 57 C 20–21 P 43, 44, 45, 47 PS 43, 44, 45, 46 R 43, 44, 45, 47

CRITERIA	LESSON/APPLICATION PAGE REFERENCES
Probability and Statistics	
6.18 The student, given a problem situation, will collect, analyze, display, and interpret data in a variety of graphical methods, including line, bar, and circle graphs and stem-and-leaf and box-and-whisker plots. Circle graphs will be limited to halves, fourths, and eighths.	218D, 234D, 236–239, 237, 242–243, 244–245, 246–247, 248, 249, 250D, 251, 252–253, 254–255, 256–257, 259, 260, 261, 262–263, 264–265, 266, 283, 286, 287, 289, 381, 479 H 59, 60, 61 C 26–27, 52–53 P 52, 54, 56, 57, 58, 59, 60 PS 54, 55, 56, 58, 109, 117, 124 R 52, 54, 56, 57, 58 E-LAB 13
6.19 The student will describe the mean, median, and mode as measures of central tendency and determine their meaning for a set of data.	228–231, 250D, 258–261, 262–263, 264–265, 266, 267, 275, 281, 283, 285, 287, 288, 305, 329, 345, 403, 455, 479, 551, 557 H 26, 41, 61 C 54–55 P 59, 60 PS 59, 60, 62, 64, 85, 86, 87, 92, 95, 100, 109, 114 R 59, 60
6.20 The student will determine and interpret the probability of an event occurring from a given sample space.	268D, 272–275, 276–277, 280–281, 282, 283, 287, 288, 293, 303, 395, 397, 403, 407, 455, 461, 479, 495, 519 H 62, 63 P 62, 63, 64 PS 62, 63, 65, 70, 76, 82, 85, 89, 92, 102, 104, 109, 113, 115, 119, 125 R 62, 63, 64
Patterns, Functions, and Algebra	
6.21 The student will recognize, describe, and extend a variety of numerical and geometric patterns.	8, 93, 375, 423, 490–491, 520D, 525, 526–527, 528–529, 530, 534, 535, 536D, 537, 538–541, 542–543, 544–545, 546–547, 548–549, 550, 551, 552, 554, 555, 556, 557 H 12, 79, 82, 83 C 98–99 P 113, 121, 122, 124, 125, 126, 127 PS 14, 122, 124, 125, 126, 127 R 121, 122, 124, 125, 126, 127 E-LAB 28
6.22 The student will investigate and describe concepts of exponents, perfect squares, and square roots, using calculators to develop the exponential patterns. Patterns will include zero and negative exponents, which lead to the idea of scientific notation. Investigations will include the binary number system as an application of exponents and patterns.	26–27, 28–29, 32, 33, 75, 78, 79, 80, 81, 91, 157, 163, 233, 455, 513 H 30, 38, 43 C 34–35, 100–101, 106–107, 114–115 P 4 PS 4, 5, 6, 8, 9, 10, 12, 14 R 4 E-LAB 1

CRITERIA	LESSON/APPLICATION PAGE REFERENCES
6.23 The student will	
model and solve algebraic equations, using concrete materials; and	298–299, 300–303, 304, 308–309, 310–313, 314–315, 318–319, 322, 326, 327, 328, 329, 339, 403, 461, 493, 499, 502–503, 535, 557 H 64 P 68, 69, 70, 72 PS 68 R 68, 69, 70, 72 E-LAB 15, 16
solve one-step linear equations in one variable, involving whole number coefficients and positive rational solutions.	298–299, 300–303, 304, 308–309, 310–313, 314–315, 318–319, 322, 326, 327, 328, 329, 339, 403, 461, 493, 499, 513, 525, 535, 557 H 64, 65 P 68, 69, 70, 72 PS 68, 69, 71, 72 R 68, 69, 70, 72 E-LAB 15, 16